Comparison of Two Regression-Based Approaches for Determining Nutrient and Sediment Fluxes and Trends in the Chesapeake Bay Watershed

By Douglas L. Moyer, Robert M. Hirsch, and Kenneth E. Hyer

Prepared in cooperation with the Virginia Department of Environmental Quality, Maryland Department of Natural Resources, and the U.S. Environmental Protection Agency Chesapeake Bay Program

Scientific Investigations Report 2012–5244

U.S. Department of the Interior
U.S. Geological Survey

U.S. Department of the Interior
KEN SALAZAR, Secretary

U.S. Geological Survey
Marcia K. McNutt, Director

U.S. Geological Survey, Reston, Virginia: 2012

For more information on the USGS—the Federal source for science about the Earth, its natural and living resources, natural hazards, and the environment, visit *http://www.usgs.gov* or call 1-888-ASK-USGS

For an overview of USGS information products, including maps, imagery, and publications, visit *http://www.usgs.gov/pubprod*

To order this and other USGS information products, visit *http://store.usgs.gov*

Suggested citation:
Moyer, D.L., Hirsch, R.M., and Hyer, K.E., 2012, Comparison of two regression-based approaches for determining nutrient and sediment fluxes and trends in the Chesapeake Bay watershed: U.S. Geological Survey Scientific Investigations Report 2012–5244, 118 p. (Available online at *http://pubs.usgs.gov/sir/2012/5244/*)

Contents

Figures

Appendix Figures

Appendix Figures—Continued

Tables

Appendix Tables

Conversion Factors and Datums

Inch/Pound to SI

Multiply	By	To obtain
Length		
mile (mi)	1.609	kilometer (km)
Area		
square mile (mi^2)	259.0	hectare (ha)
square mile (mi^2)	2.590	square kilometer (km^2)
Volume		
cubic foot (ft^3)	0.02832	cubic meter (m^3)
Flow rate		
cubic foot per second (ft^3/s)	0.02832	cubic meter per second (m^3/s)
cubic foot per second per square mile [(ft^3/s)/mi^2]	0.01093	cubic meter per second per square kilometer [(m^3/s)/km^2]
Mass		
ton, short (2,000 lb)	0.9072	megagram (Mg)
ton per day (ton/d)	0.9072	metric ton per day
ton per day per square mile [(ton/d)/mi^2]	0.3503	megagram per day per square kilometer [(Mg/d)/km^2]
ton per year (ton/yr)	0.9072	megagram per year (Mg/yr)

Vertical coordinate information is referenced to the North American Vertical Datum of 1988 (NAVD 88).

Horizontal coordinate information is referenced to the North American Datum of 1983 (NAD 83).

Abbreviations

AMLE	Adjusted Maximum Likelihood Estimation
BNR	biological nitrate reduction
CBP	Chesapeake Bay Program
ESTIMATOR	a multiple regression model
$(lb/d)/mi^2$	pounds per day per square mile
MDDNR	Maryland Department of Natural Resources
mg/L	milligram per liter
NO_2	nitrite
NO_3	nitrate
PO_4^{3-}	orthophosphorus
RIM	River Input Monitoring
SSC	suspended-sediment concentration
TMDL	total maximum daily load
TSS	total suspended solids
USEPA	U.S. Environmental Protection Agency
USGS	U.S. Geological Survey
VADEQ	Virginia Department of Environmental Quality
WRTDS	Weighted Regressions on Time, Discharge, and Season

Comparison of Two Regression-Based Approaches for Determining Nutrient and Sediment Fluxes and Trends in the Chesapeake Bay Watershed

By Douglas L. Moyer, Robert M. Hirsch, and Kenneth E. Hyer

Abstract

Nutrient and sediment fluxes and changes in fluxes over time are key indicators that water resource managers can use to assess the progress being made in improving the structure and function of the Chesapeake Bay ecosystem. The U.S. Geological Survey collects annual nutrient (nitrogen and phosphorus) and sediment flux data and computes trends that describe the extent to which water-quality conditions are changing within the major Chesapeake Bay tributaries. Two regression-based approaches were compared for estimating annual nutrient and sediment fluxes and for characterizing how these annual fluxes are changing over time. The two regression models compared are the traditionally used ESTIMATOR and the newly developed Weighted Regression on Time, Discharge, and Season (WRTDS). The model comparison focused on answering three questions: (1) What are the differences between the functional form and construction of each model? (2) Which model produces estimates of flux with the greatest accuracy and least amount of bias? (3) How different would the historical estimates of annual flux be if WRTDS had been used instead of ESTIMATOR? One additional point of comparison between the two models is how each model determines trends in annual flux once the year-to-year variations in discharge have been determined. All comparisons were made using total nitrogen, nitrate, total phosphorus, orthophosphorus, and suspended-sediment concentration data collected at the nine U.S. Geological Survey River Input Monitoring stations located on the Susquehanna, Potomac, James, Rappahannock, Appomattox, Pamunkey, Mattaponi, Patuxent, and Choptank Rivers in the Chesapeake Bay watershed.

Two model characteristics that uniquely distinguish ESTIMATOR and WRTDS are the fundamental model form and the determination of model coefficients. ESTIMATOR and WRTDS both predict water-quality constituent concentration by developing a linear relation between the natural logarithm of observed constituent concentration and three explanatory variables—the natural log of discharge, time, and season. ESTIMATOR uses two additional explanatory variables—the square of the log of discharge and time-squared. Both models determine coefficients for variables for a series of estimation windows. ESTIMATOR establishes variable coefficients for a series of 9-year moving windows; all observed constituent concentration data within the 9-year window are used to establish each coefficient. Conversely, WRTDS establishes variable coefficients for each combination of discharge and time using only observed concentration data that are similar in time, season, and discharge to the day being estimated. As a result of these distinguishing characteristics, ESTIMATOR reproduces concentration-discharge relations that are closely approximated by a quadratic or linear function with respect to both the log of discharge and time. Conversely, the linear model form of WRTDS coupled with extensive model windowing for each combination of discharge and time allows WRTDS to reproduce observed concentration-discharge relations that are more sinuous in form.

Another distinction between ESTIMATOR and WRTDS is the reporting of uncertainty associated with the model estimates of flux and trend. ESTIMATOR quantifies the standard error of prediction associated with the determination of flux and trends. The standard error of prediction enables the determination of the 95-percent confidence intervals for flux and trend as well as the ability to test whether the reported trend is significantly different from zero (where zero equals no trend). Conversely, WRTDS is unable to propagate error through the many (over 5,000) models for unique combinations of flow and time to determine a total standard error. As a result, WRTDS flux estimates are not reported with confidence intervals and a level of significance is not determined for flow-normalized fluxes.

The differences between ESTIMATOR and WRTDS, with regard to model form and determination of model coefficients, have an influence on the determination of nutrient and sediment fluxes and associated changes in flux over time as a result of management activities. The comparison between the model estimates of flux and trend was made for combinations of five water-quality constituents at nine River Input Monitoring stations.

The major findings with regard to nutrient and sediment fluxes are as follows: (1) WRTDS produced estimates of flux for all combinations that were more accurate, based on reduction in root mean squared error, than flux estimates from ESTIMATOR; (2) for 67 percent of the combinations, WRTDS and ESTIMATOR both produced estimates of flux that were minimally biased compared to observed fluxes (flux bias=tendency to over or underpredict flux observations); however, for 33 percent of the combinations, WRTDS produced estimates of flux that were considerably less biased (by at least 10 percent) than flux estimates from ESTIMATOR; (3) the average percent difference in annual fluxes generated by ESTIMATOR and WRTDS was less than 10 percent at 80 percent of the combinations; and (4) the greatest differences related to flux bias and annual fluxes all occurred for combinations where the pattern in observed concentration-discharge relation was sinuous (two points of inflection) rather than linear or quadratic (zero or one point of inflection).

The major findings with regard to trends are as follows: (1) both models produce water-quality trends that have factored in the year-to-year variations in flow; (2) trends in water-quality condition are represented by ESTIMATOR as a trend in flow-adjusted concentration and by WRTDS as a flow-normalized flux; (3) for 67 percent of the combinations with trend estimates, the WRTDS trends in flow-normalized flux are in the same direction and magnitude to the ESTIMATOR trends in flow-adjusted concentration, and at the remaining 33 percent the differences in trend magnitude and direction are related to fundamental differences between concentration and flux; and (4) the majority (85 percent) of the total nitrogen, nitrate, and orthophosphorus combinations exhibited long-term (1985 to 2010) trends in WRTDS flow-normalized flux that indicate improvement or reduction in associated flux and the majority (83 percent) of the total phosphorus (from 1985 to 2010) and suspended sediment (from 2001 to 2010) combinations exhibited trends in WRTDS flow-normalized flux that indicate degradation or increases in the flux delivered.

Introduction

Excessive nutrient (nitrogen and phosphorus) and sediment transport to the Chesapeake Bay from the watershed is detrimental to the overall structure and function of the bay ecosystem and is a major concern for local, State, and Federal entities that benefit from and work to protect the living resources of the bay. The flux (also called load) of nutrients to the Chesapeake Bay is in part natural but has been accelerated as a result of anthropogenic inputs of these nutrients through sewage disposal, agricultural runoff, urban runoff, and acid rain (Officer and others, 1984; Nixon, 1987; Schlesinger, 1997). Accelerated eutrophication through excessive nutrient flux has been linked to the loss of critical habitat for living resources within the Chesapeake Bay estuary (U.S. Environmental Protection Agency, 1983). Cooper and Brush (1991) found that accelerated algal production resulting from elevated nutrient fluxes has led to an increased occurrence of anoxic conditions in bottom waters and associated sediment throughout the Chesapeake Bay estuary. Similarly, the flux of sediment to the Chesapeake Bay results from both natural processes associated with upland erosion, lateral movement of channels into streambanks, and downcutting of streambeds (Waters, 1995) as well as anthropogenic processes such as agriculture, logging, mining, and urbanization. Anthropogenically derived sediment can overwhelm the natural assimilative capacity of the aquatic system (Cairns, 1977) and may bury filter-feeding organisms, reduce habitat available for macroinvertebrates, contribute to decreased fish populations, and impair growth of aquatic vegetation by reducing available light (Lenat and others, 1981; Dennison and others, 1993; Box and Mossa, 1999; Madsen and others, 2001).

The Chesapeake Bay Program (CBP) was initiated in 1983 to direct the restoration and protection of the Chesapeake Bay. The CBP is composed of various Federal, State, academic, and local watershed organizations. In 1987, the CBP established its first nutrient reduction goal, which was to reduce nitrogen and phosphorus fluxes to the Chesapeake Bay. In 2000, the CBP recommitted to achieve the nutrient and sediment reduction goals established in 1987and established criteria for dissolved oxygen, chlorophyll, and water clarity (Chesapeake Bay Program, 2000). Despite extensive restoration efforts made by the CBP, however, established water-quality goals were not being obtained for the Chesapeake Bay and associated tributaries (Chesapeake Bay Foundation, 2010). As a result, in 2010, the U.S. Environmental Protection Agency (USEPA) established the Chesapeake Bay total maximum daily load (TMDL) for nitrogen, phosphorus, and sediment (U.S. Environmental Protection Agency, 2010). This TMDL assigns accountability for nutrient and sediment fluxes to New York, Pennsylvania, Maryland, Delaware, Virginia, West Virginia, and the District of Columbia and serves as a catalyst for rigorous implementation of management actions to mitigate the transport of excessive nutrients and sediment to the Chesapeake Bay and tidal estuaries.

Since the early 1990s, the U.S. Geological Survey (USGS), in cooperation with the Virginia Department of Environmental Quality (VADEQ) and the Maryland Department of Natural Resources (MDDNR), has been responsible for monitoring nutrient and sediment conditions in the major rivers of the Chesapeake Bay watershed. Additionally, the USGS is responsible for quantifying annual nutrient and sediment fluxes to the bay as well as determining long-term changes in water-quality conditions in the Chesapeake Bay watershed to facilitate continuing evaluation of the progress being made toward reducing nutrient and sediment inputs to the bay. The primary method for quantifying fluxes and determining trends in concentration at the River Input Monitoring (RIM) stations has been through the use of a multiple regression model (ESTIMATOR). ESTIMATOR had its origins in the work of Cohn and others (1992) who determined that the multiple regression approach generates valid flux estimates for nutrients (suspended sediment was not evaluated); the evaluation was based on data collected between 1980 and 1988 at the four RIM stations in Maryland. The model was modified to accommodate censored values (water-quality concentrations reported as less than a specified reporting or analytical limit) (Cohn, 2005). ESTIMATOR has been widely used for flux determination and identification of changing water-quality conditions over time (Langland and others, 2006; Aulenbach and others, 2007). As the USGS RIM program generated water-quality datasets for longer periods of time, however, it became apparent that improvements needed to be made to the ESTIMATOR approach. These improvements included (1) enhanced model flexibility to accommodate complex concentration-discharge relations (that is, concentration-discharge relations that are not linear or quadratic in form and that can change over a period of several decades) and (2) the ability to provide estimates of trends in flux, which may be different from trends in concentration (that is, an inherent property of ESTIMATOR is that, when expressed in terms of percentage change over time, estimates of concentration changes are constrained to be equal to percentage changes in flux). An evolving understanding of these apparent shortfalls in the ESTIMATOR approach has led to research and development of new methods for the determination of nutrient and sediment fluxes and associated trends at long-term (greater than 20 years) monitoring stations with large (greater than 300 observations) water-quality datasets.

To address these needs, the USGS recently developed a new method for the determination of nutrient and sediment fluxes and trends using multiple weighted regressions (Hirsch and others, 2010). The Weighted Regressions on Time, Discharge, and Season (WRTDS) method was developed to provide a more robust tool for quantifying concentrations, fluxes, and descriptions of long-term changes in these quantities at monitoring stations with long-term datasets. In particular, it is designed to provide these descriptions of long-term changes in a manner that is not influenced by the particular year-to-year variations in river discharge, but rather provides a description of the evolving nature of the overall behavior of the watershed system in terms of nutrient and sediment concentrations and fluxes. This technique produces results that are directly relevant to the needs of the CBP by describing estimates of the yearly or seasonal nutrient and sediment inputs to the bay, as well as providing insight into the effects land-management actions have on water-quality

conditions in the major tributaries to the bay. In addition to its use with Chesapeake Bay RIM data (Hirsch and others, 2010), the WRTDS method also has been applied to studies of long-term changes in the Mississippi River Basin (Sprague and others, 2011) and in the Lake Champlain Basin (Medalie and others, 2012).

In 2011, the USGS began an investigation to compare the flux and trend estimates derived from ESTIMATOR and WRTDS. The overall objective of the investigation was to evaluate the nature and extent of the differences between nutrient and sediment flux estimates generated by each method and to determine which model provides the highest level of accuracy in annual flux estimates provided to the Chesapeake Bay Program. The model comparison focused on answering three questions: (1) What are the differences between the functional form and construction of the two models? (2) Which model produces discrete daily estimates of flux with the greatest accuracy and least amount of bias? and (3) How different would the historical estimates of annual flux be if WRTDS had been used instead of ESTIMATOR? One additional point of comparison between the two models was how each model determines the changes in annual flux once the year-to-year variations in discharge have been accounted for. All comparisons were made using total nitrogen, nitrate, total phosphorus, orthophosphorus, and suspended-sediment concentration data collected at the nine USGS RIM stations located on the Susquehanna, Potomac, James, Rappahannock, Appomattox, Pamunkey, Mattaponi, Patuxent, and Choptank Rivers in the Chesapeake Bay watershed. The investigation provided valuable information that extends beyond the boundaries of the Chesapeake Bay watershed with regard to determining the appropriate regression-based approach for quantifying nutrient and sediment fluxes and identifying how these fluxes are changing over time.

Purpose and Scope

This report documents the comparison of two multiple regression approaches, ESTIMATOR and WRTDS, for the determination of nutrient and sediment fluxes in the Chesapeake Bay watershed. Discharge, nutrient, and sediment data collected from 1985 through 2010 at the RIM stations on the Susquehanna, Potomac, James, Rappahannock, Appomattox, Pamunkey, Mattaponi, Patuxent, and Choptank Rivers were used to construct ESTIMATOR and WRTDS models for predicting daily constituent concentrations and fluxes. ESTIMATOR and WRTDS discrete flux estimates were compared to direct observations of flux at each of the nine RIM stations to determine the accuracy and bias associated with each model. Annual estimates of nutrient and sediment flux were compared to determine the average difference in annual fluxes generated by ESTIMATOR and WRTDS. Finally, this report provides trends in WRTDS-derived annual nutrient and sediment fluxes, represented as the flow-normalized annual flux, for the periods 1985–2010 and 2001–2010.

Description of Nutrient and Sediment Data

The USGS monitors nutrient and sediment conditions in the Chesapeake Bay watershed at nine long-term RIM stations (fig. 1; table 1). These RIM stations are situated at the furthest downstream point on the river, prior to where the river becomes tidally influenced. Because of historical data and logistical issues, however, the James River RIM station is located approximately 45 miles upstream from the tidal influence. The nine RIM stations, combined, account for streamflow from approximately 78 percent of the land area in the entire watershed (Langland and others, 1995). Monitoring water-quality conditions at these RIM stations allows for a nearly comprehensive representation of the total flux delivered to the tidal estuaries of the Chesapeake Bay from the nontidal portion of the watershed.

Since the early 1980s, the USGS has collected a minimum of 20 samples per year at each of the nine RIM stations. These samples are collected across the full range of the hydrologic conditions and are composed of 12 monthly samples and 8 targeted stormflow (that is, periods of elevated discharge) samples. These samples are analyzed for a variety of constituents that include dissolved and particulate phases of nitrogen and phosphorus as well as suspended sediment.

For the model comparison, the following constituents were compiled into datasets for use in ESTIMATOR and WRTDS: total nitrogen, nitrate-nitrogen (analyzed as the mass of nitrogen in both nitrite+nitrate ($NO_2^- + NO_3^-$)) (referred to hereafter as nitrate), total phosphorus, orthophosphorus (PO_4^{3-}), and suspended sediment. These water-quality data exist as either measured or calculated values. Measured values are those that are directly quantified through laboratory analysis; calculated values are those that are determined as the sum of a set of associated measured constituents. The priority, in constructing water-quality datasets as input for both ESTIMATOR and WRTDS, was to use a measured value whenever possible before using a calculated value. Nitrate, orthophosphorus, and suspended sediment are only available as measured values; however, historical values of total nitrogen and total phosphorus exist in some cases as both measured and calculated. Therefore, if there are concurrent data, the order of operation for constructing the time series datasets for total nitrogen, in order of decreasing priority, is as follows:

1. Total nitrogen=total nitrogen (measured),

2. Total nitrogen=total dissolved nitrogen +total particulate nitrogen, or

3. Total nitrogen=total Kjeldahl nitrogen+nitrate.

Table 1.   Chesapeake Bay River Input Monitoring Stations and associated watershed characteristics.

[mi^2, square mile; TN, total nitrogen; NO$_3$, nitrate; TP, total phosphorus; OP, orthophosphorus; SSC, suspended sediment]

Station name	Station number	Drainage area (mi^2)	Number of samples collected during 1985 to 2010					Site abbreviation	Map identifier
			TN	NO$_3$	TP	OP	SSC		
Susquehanna River at Conowingo, Md.	01578310	27,100	904	909	906	906	901	SUS	1
Potomac River at Chain Bridge at Washington, D.C.	01646580	11,600	1,427	1,429	1,432	1,361	475	POT	2
James River at Cartersville, Va.	02035000	6,252	769	784	772	780	245	JAM	3
Rappahannock River near Fredericksburg, Va.	01668000	1,595	695	711	702	705	225	RAP	4
Appomattox River at Matoaca, Va.	02041650	1,342	714	731	721	725	246	APP	5
Pamunkey River near Hanover, Va.	01673000	1,078	758	783	769	776	235	PAM	6
Mattaponi River near Beulahville, Va.	01674500	603	746	765	756	767	221	MAT	7
Patuxent River near Bowie, Md.	01594440	348	786	874	864	843	850	PAT	8
Choptank River near Greensboro, Md.	01491000	113	622	624	616	612	684	CHO	9

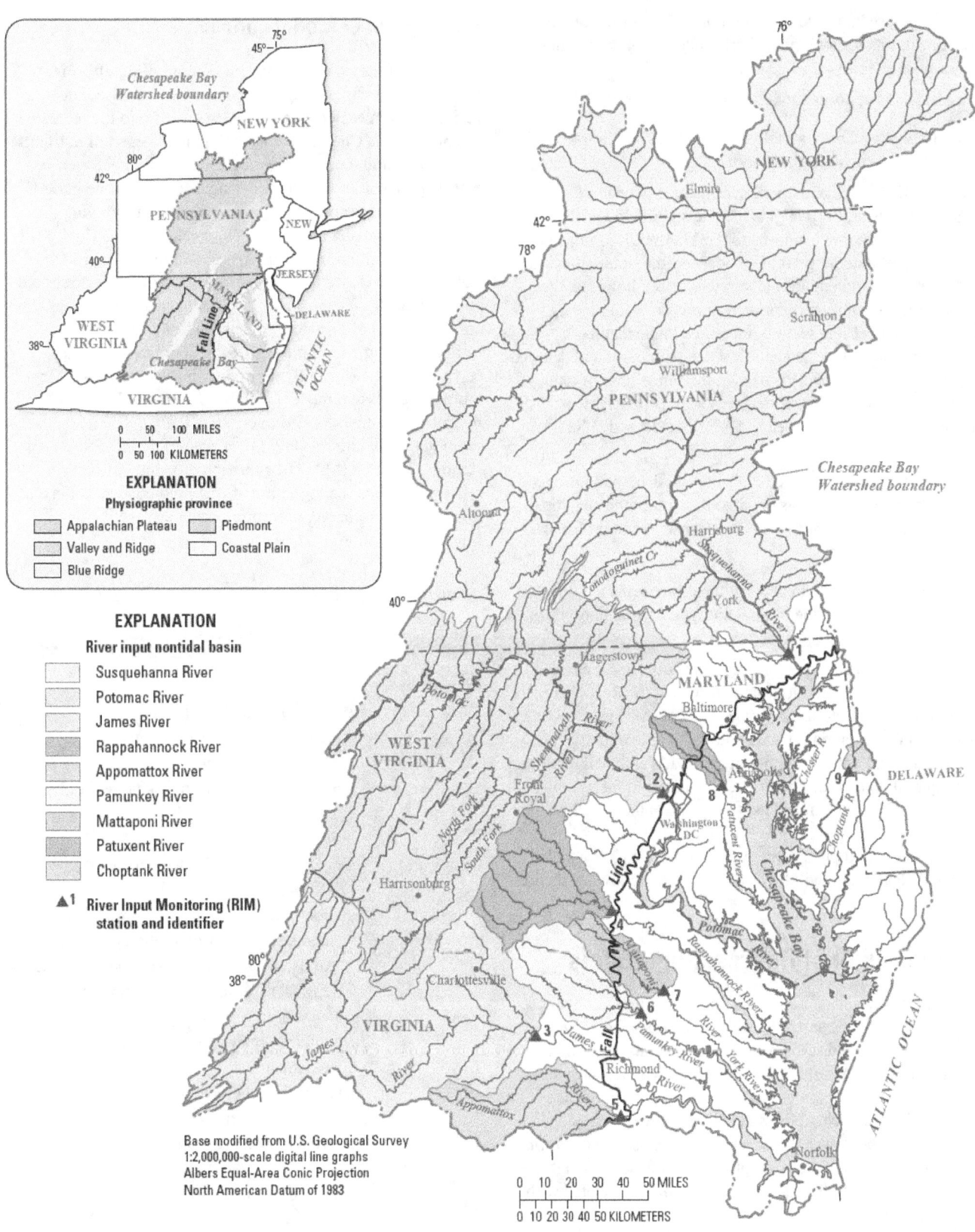

Figure 1. Location of the nine River Input Monitoring (RIM) stations in the Chesapeake Bay watershed. Station numbers and names are provided in table 1.

Likewise, the order of operation for constructing the time series datasets for total phosphorus, in order of decreasing priority, is as follows:

1. Total phosphorus = total phosphorus (measured) or

2. Total phosphorus = total dissolved phosphorus + total particulate phosphorus.

A distinction in the construction of these input datasets for ESTIMATOR and WRTDS is that the summation of total nitrogen and total phosphorus for ESTIMATOR input datasets is performed externally to the model and the summations for WRTDS are handled internally to the model; however, the same order of summation priority applies to both models.

When a censored value (measured value that has a concentration below the analytical detection limit and denoted by a "less than" symbol) is used in the summation of total nitrogen and total phosphorus, censoring is propagated differently for datasets used in ESTIMATOR and WRTDS. For ESTIMATOR,

1. if there are no censored constituents, then total nitrogen or total phosphorus is equal to the sum of the analytical constituents according to the order of operation listed above; however,

2. if any of the analytical constituents are censored, then the final summed value of total nitrogen or total phosphorus is censored .

An example of censoring propagation shown in (2) can be found in the summation of total nitrogen at the Rappahannock RIM station on October 10, 2007. The constituents available for the summation of total nitrogen are total particulate nitrogen (<0.03 milligram per liter (mg/L)) and total dissolved nitrogen (0.28 mg/L). Therefore, total nitrogen for use in ESTIMATOR is calculated as <0.03 mg/L + 0.28 mg/L, which is equal to <0.31 mg/L. In WRTDS, the summation of total nitrogen and total phosphorus with censored data is handled within WRTDS and will be discussed in the following section.

Comparison of Regression Models

An important step in comparing model results obtained from ESTIMATOR and WRTDS is to compare how the models are used to estimate water-quality constituent flux at the nine RIM stations. This section of the report provides information that compares (1) the functional form of each model, (2) how each model's coefficients are estimated, data are handled, (4) how these models are used to determine temporal changes in the concentration and flux, and (5) the ability of each model to quantify uncertainty associated with flux estimates.

Description of Model Forms

The primary goal for the USGS monitoring effort is to quantify the flux of nitrogen, phosphorus, and sediment delivered from each of the nine RIM basins into the receiving estuaries of the Chesapeake Bay. Flux is the mass of nutrients and sediment that passes the monitoring station before entering the Chesapeake Bay during a specified time period. A daily flux (mass per day) is calculated by multiplying the daily streamflow by the constituent concentration. The USGS continuously monitors streamflow conditions at each RIM station; however, water-quality conditions are only monitored approximately 20 days during a given year. Therefore, daily concentrations of nitrogen, phosphorus, and sediment are estimated using a multiple regression approach. Since the early 1990s, the USGS RIM program has used a log-linear multiple regression model (ESTIMATOR) developed by Cohn and others (1989) to estimate daily nutrient and sediment concentration and flux at each of the nine RIM stations. The ESTIMATOR approach produces a best fit relation between the logarithm of observed concentration and logarithm of discharge, time, and season as described in the following equation:

$$\ln(c) = \hat{\beta}_0 + \hat{\beta}_1 \ln(q/q_c) + \hat{\beta}_2 \left[\ln(q/q_c)\right]^2 + \hat{\beta}_3(t-t_c)$$
$$+ \hat{\beta}_4(t-t_c)^2 + \hat{\beta}_5 \sin(2\pi t) + \hat{\beta}_6 \cos(2\pi t) + \varepsilon, \qquad (1)$$

where

$\ln$	is the natural log function;
c	is the measured concentration, in milligrams per liter;
q	is measured daily-mean discharge, in cubic feet per second;
t	is time, in decimal years;
q_c, t_c	are centering variables for streamflow and time;
$\hat{\beta}$	are coefficients estimated by ordinary least squares (non-censored observation) and Adjusted Maximum Likelihood Estimation (AMLE) (censored observations);
$\hat{\beta}_0$	is a constant;
$\hat{\beta}_1, \hat{\beta}_2$	describe the relation between concentration and streamflow;
$\hat{\beta}_3, \hat{\beta}_4$	describe the relation between concentration and time, independent of flow;
$\hat{\beta}_5, \hat{\beta}_6$	describe seasonal variation in concentration; and
ε	is the unexplained variation.

ESTIMATOR predicts constituent concentration in log space using seven explanatory variables. Cohn and others (1992) demonstrated that for several datasets considered (Chesapeake Bay RIM stations in Maryland, for nitrogen and phosphorus constituents only), this model accounts for a substantial portion of the variation in concentration and results in residuals that are approximately homoscedastic (unchanging variability) and not highly correlated with any of the predictor variables. Variation in concentration as a function of discharge is addressed by including both discharge and discharge-squared terms; these terms allow predicted concentration to change linearly or parabolically as a function of streamflow. Therefore, ESTIMATOR does well with modeling observed concentration-discharge relations that are linear or quadratic in form. Variation in concentration as a function of time is accounted for by including both time and time-squared; these variables are used to remove linear and (or) parabolic trends in the concentration residuals resulting from long-term increase/decrease in concentration. Variations in concentration as a function of season is addressed by including sine and cosine functions with period $2\pi t$; variable coefficients on these terms, determined as part of model estimation, allow the model to account for a single sinusoidal cycle with any magnitude or phase. In addition to the seven explanatory variables, ESTIMATOR uses centering variables for flow (q_c) and time (t_c) to ensure that time and time-squared and discharge and discharge-squared are orthogonal (independent). These centering variables simplify calculations in ESTIMATOR and have no effect on flux estimates (Cohn and others, 1992). There is an implicit assumption in ESTIMATOR that the shape of the relation between concentration and flow is constant throughout the estimation period although the intercept is free to vary as a quadratic function of time. Similarly, there is an implicit assumption that the seasonality of this flow versus concentration relation can be described as a sine wave of fixed amplitude and phase throughout the estimation period.

The USGS is now considering adding the WRTDS approach, developed by Hirsch and others (2010), for the determination of nutrient and sediment fluxes and associated trends at each of the nine RIM stations to ensure the integrity of flux and trend estimates and to help overcome the potential limitations associated with ESTIMATOR. The WRTDS approach is described in detail by Hirsch and others (2010) and Sprague and others (2011). Like ESTIMATOR, WRTDS produces a best-fit relation defined between the logarithm of observed concentration and logarithm of discharge, time, and season as described in the following equation:

$$\ln(c) = \hat{\beta}_0 + \hat{\beta}_1 t + \hat{\beta}_2 \ln(q) + \hat{\beta}_3 \sin(2\pi t) + \hat{\beta}_4 \cos(2\pi t) + \varepsilon$$

(2)

where

ln	is the natural log function;
c	is the measured concentration, in milligrams per liter;
q	is measured daily-mean discharge, in cubic feet per second;
t	is time, in years;
$\hat{\beta}$	are coefficients;
$\hat{\beta}_0$	is a constant;
$\hat{\beta}_1$	describes the relation between concentration and time;
$\hat{\beta}_2$	describes the relation between concentration and flow;
$\hat{\beta}_3, \hat{\beta}_4$	describe seasonal variation in concentration; and
ε	is the unexplained variation.

WRTDS is similar in functional form to ESTIMATOR in that it uses discharge, time, and season as variables to explain the variation associated with observed water-quality constituents. WRTDS, however, explains this variation with five explanatory variables compared to the seven variables within ESTIMATOR. WRTDS does not include discharge-squared and time-squared variables and as a result does not need centering variables for time and discharge to ensure orthogonality. The key difference between the two methods is the way that the coefficients vary as a function of time, discharge, and season.

Parameter Estimation

Parameter estimation refers to the process by which model parameters/coefficients are estimated to determine constituent concentration and flux. Two key steps in this process that distinguishes ESTIMATOR from WRTDS is the width of the model estimation window and the means by which observed water-quality data are incorporated in the estimation window. The model construction processes used in this investigation will be described for ESTIMATOR and WRTDS.

The parameter estimation approach that the USGS currently (2012) uses for the determination of constituent concentration and flux, using ESTIMATOR, begins with the establishment of the model window. Until 2000, a single model window was used that encompassed the entire period of the water-quality monitoring record (for example, 1979 to 1999). This approach employed two assumptions: (1) time invariance in the relation between concentration and discharge, time, and season and (2) equal uncertainty associated with all annual estimates of flux within the full estimation window. These assumptions held true when the model estimation window was relatively short (<10 years); however, as the monitoring program matured and the datasets expanded to include 15 or more years of data, these assumptions became less valid (Yochum, 2000; Milly, 2008). As the dataset grew, the model fit became overly sensitive to data from the beginning and ending years. Particularly with the use of the time-squared variable, errors could become substantial, especially if the trends in the data did not conform to the quadratic or linear shape (Yochum, 2000). Yochum (2000) recommended the use of a series of 9-year estimation windows when using ESTIMATOR to quantify monthly and annual fluxes. This approach minimizes estimation error primarily by allowing the relation between concentration and discharge, time, and season flexibility to change over time. Yochum (2000) showed that to produce fluxes with the greatest accuracy, a 9-year estimation window should be used to estimate fluxes for the centered (5th) year only and to estimate fluxes for the entire period a 9-year moving window approach should be used.

Since 2000, the USGS has used a 9-year moving window approach to estimate monthly and annual fluxes at the nine RIM stations. All available observations for the constituent being modeled, for each 9-year window, are used to estimate the seven model coefficients in equation 1, including those for flow, time, and season through best-fit ordinary least squares. Once the seven model coefficients are defined, daily concentrations are calculated from daily discharge, time, and season. ESTIMATOR uses the AMLE developed by Cohn (2005) to (1) estimate daily concentrations when censored constituent data are present and (2) account for retransformation bias that occurs when the natural log of estimated concentrations are retransformed from log space. Finally, the constituent flux is determined for each day by multiplying the daily concentration by the daily discharge. Monthly and annual fluxes, for the centered year in the 9-year window (beginning with 1985 which is centered in the 1981 to 1989 9-year window), are the summation of the daily fluxes for the associated month or year, respectively. Fluxes for the most recent 4 years are obtained from the last 4 years of the final 9-year estimation window and labeled as provisional (Yochum, 2000). As an example, the current investigation obtained annual fluxes for 2007, 2008, 2009, and 2010 from the ESTIMATOR 9-year window spanning 2002 to 2010.

In contrast, WRTDS uses a considerably different approach than ESTIMATOR for the determination of nutrient and sediment fluxes. The first major distinction is that instead of the 9-year estimation window used by ESTIMATOR,

unique WRTDS models (eq. 2) are constructed for a large number of combinations of discharge (Q) and time (T) defined over a grid. The first dimension of the grid is time. The number of values of T is set to (years $\times$ 16)+1, where years is the length of the period for which estimates are being made. Therefore, the grid spacing is set to 1/16 of a year (approximately every 23 days). The second dimension in the estimation grid is discharge, where discharge has 14 equally spaced (in log of discharge) levels that range from slightly higher than the highest daily discharge to slightly lower than the lowest daily discharge. At each unique grid intersection between discharge and time, unique WRTDS models are constructed, and estimated values of concentration are stored. The total number of unique WRTDS models is $14 \times ((years \times 16)+1)$; therefore, for each water constituent modeled, as part of this report, there are $14 \times ((26 \times 16)+1)$ WRTDS models or 5,838 unique WRTDS models. WRTDS uses bilinear interpolation to estimate concentration for combinations of Q and T that do not coincide with the estimation grid nodes.

Another factor that distinguishes WRTDS from ESTIMATOR is the method by which water-quality observations are chosen for the estimation of model coefficients. WRTDS identifies which water-quality observations are included in each unique model by weighting each observation in the entire dataset on the basis of similarities/distance from the target condition in three dimensions: time, season, and discharge (Hirsch and others, 2010). The first distance assessed is the time distance where greater weight is given to observations that were collected closer in time to the target time. For this investigation, the half-window width is set to 10 years; therefore, observations that are approximately 6 years from the center of the window are assigned weights that are less than half of the weights at the center of the window. Observations 10 years and greater from the center of the window are assigned weights equal to zero. The second distance is the seasonal distance where greater weight is given to observations that were collected during the same time of year. The half-window width, for this investigation, is set to 0.5 meaning, for example, if July 1, 2010, is the unique time where a WRTDS model is being constructed, then observations collected during the summer would have the greatest weights followed by observations collected during spring and fall; observations collected a half a year from the time being estimated (in this example January 1) would be assigned weights equal to zero. The third distance is the discharge distance where greater weight is given to observations collected during similar discharge conditions/magnitude. The half-window width for discharge (log discharge), for this investigation, is set to 2; therefore, water-quality observations that were collected during discharge conditions that are within two natural log cycles of the target discharge condition will be weighted greater than zero and included in the model coefficient estimation process. The "tri-cubed weight function" (Tukey, 1977) is defined in each of these three dimensions, and then these three weights are combined by multiplying these together to determine the overall weight of the observation. The greater the overall weight of an observation, the greater the influence it has in establishing the parameters in equation 2. A minimum of 100 observations with weights greater than zero is required in each model window. Hirsch and others (2010) provide extensive details pertaining to the how weights are assigned to observations. The coefficients in equation 2 are fitted, using the weighted observations, and used to estimate constituent concentration that best represents the targeted discharge and time. Retransformation bias associated with transforming concentration back from log space was addressed in the version of WRTDS used by Hirsch and others (2010) by using the smearing factor developed by Duan (1983); however, the version of WRTDS used for this investigation (version 4) uses a method that addresses retransformation bias and fitting of model coefficients in the presence of censored data simultaneously and is discussed in detail in the section "Censored Data." Following this approach, WRTDS produces a concentration derived from unique models of individual flow and time for every day in the monitoring record. Daily flux is then determined by multiplying the daily concentration by the daily discharge. Daily fluxes are summed to obtain monthly and annual fluxes.

The estimation of model coefficients for ESTIMATOR and WRTDS is similar in that they both use the concept of a moving window so that the regression model is based on observations that are from a period of years "near" to the year for which the estimates are being computed. One difference is that the windowed ESTIMATOR approach treats observations as either "in" or "out" of the regression, so that at each annual increment of time individual observations are added to the regression dataset or deleted. In contrast, in WRTDS the window moves continuously through time rather than moving forward a year at a time. The portion of the total weight attributed to time on each observation changes gradually from zero to one (versus an abrupt shift from zero to one and back to zero again). As a consequence, in WRTDS the influence of each observation on the regression gradually changes with the passage of time. The other difference is that ESTIMATOR creates a window only in the time dimension, whereas WRTDS uses additional windows in discharge and season. Thus, in ESTIMATOR data from all discharge values and seasons of the year are given equal weight in the regression as long as they are in the time window. In WRTDS, the weighting scheme provides high weights for data with similar time, discharge, and season when compared to the target condition, and low weights are provided for data with dissimilar time, discharge, and season when compared to the target condition. Because of this fundamental difference, WRTDS theoretically replicates the behavior of the constituents at higher discharges where data tend to be relatively sparse, but which are highly important in the overall annual flux. In a sense, the estimates for the high discharges are decoupled from the behavior observed at much lower discharges, which may depend on very different factors and processes.

Censored Data

The ability of ESTIMATOR and WRTDS to determine daily concentrations and fluxes for constituents that have censored data observations is essential. ESTIMATOR uses the AMLE methodology to handle the estimation of constituent concentration and flux in the presence of censored data; details are given in Cohn (2005). The version of WRTDS used in this investigation (version 4) is different from the version reported in Hirsch and others (2010) because version 4 allows for computation of estimated concentration values in the presence of censored values. The technique is an adaptation of "survival analysis," originally developed for medical or industrial applications, which is also known as "censored regression analysis." WRTDS (version 4) allows for left censoring and interval censoring of the observed concentration data. Left censoring is a common characteristic of water-quality data (Cohn, 2005; Helsel, 2012). Interval censoring occurs when the concentration of interest is the sum of two or more concentration values and at least one is reported as a "less than" value and at least one is reported without censoring; this situation occurs frequently in the RIM data. For example, for the Rappahannock River RIM station, 41 of 695 samples are reported for total nitrogen for which there is interval censoring (and none with left censoring). One of these instances is the sample taken on September 11, 2003, where total dissolved nitrogen was reported as 0.64 mg/L and total particulate nitrogen was reported was <0.01 mg/L. Because the analysis for which equation 2 is being fit is for total nitrogen, these two results must be added together to constitute an estimate of total nitrogen concentration. Given this information, the value of total nitrogen lies between 0.64 mg/L and 0.65 mg/L. Note that if the analysis was restricted to only using left censoring representations of results, as with ESTIMATOR, a value of <0.65 mg/L would be assigned to this observation, which conveys very different information. Figure 2 illustrates the representation of all 695 values in this dataset, using the vertical lines to indicate the range of each interval estimate.

The general rule, in WRTDS, for computing interval estimates between the concentration of analyte 1 (c_1) and the concentration of analyte 2 (c_2) is accomplished by implementing 1 of 3 possible cases. The reporting limits for the two analytes are r_1 and r_2, respectively. The method computes two values: I_L, which is the lower limit of the range of possible values for the sum, and I_U, which is the upper limit of the range. If c_1 and c_2 are not censored, then I_L and I_U are identical. For illustrative purposes, it is assumed that there are never more than two analytes present for the summation of total nitrogen and total phosphorus; however, this interval-censored approach is generalized in the WRTDS software to apply to any number of analytes. The three interval censoring cases are

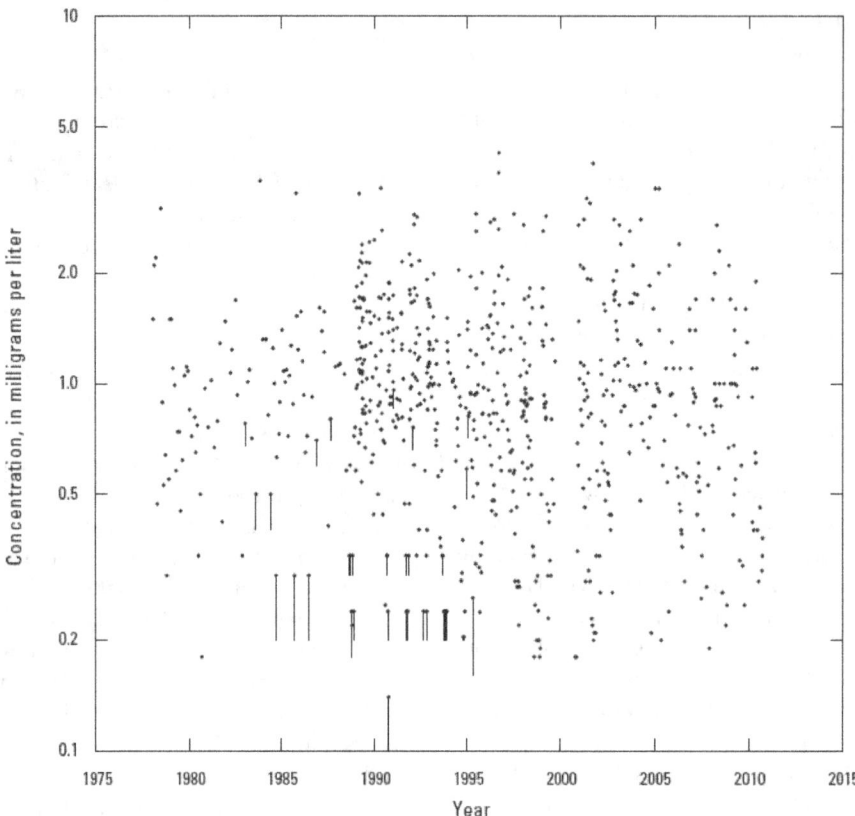

Figure 2.   Total nitrogen record, showing interval censored values as vertical lines, for Rappahannock River at Fredericksburg, Virginia.

- Case 1: the concentration of analytes 1 and 2 are greater than or equal to their reporting limits, that is $c_1 \geq r_1$ and $c_2 \geq r_2$. In this case, the concentration assigned to the lower (I_L) and upper (I_U) limits of the concentration interval is defined as $I_L = c_1 + c_2$ and $I_U = c_1 + c_2$. This means that the interval estimate collapses to a single value, $c_1 + c_2$.

- Case 2: the concentration of analyte 1 is reported as less than the reporting limit, but the concentration of analyte 2 is reported as greater than or equal to its reporting limit, that is $c_1 < r_1$ and $c_2 \geq r_2$. In this case, $I_L = c_2$ and $I_U = r_1 + c_2$. This case was illustrated above with the example using total nitrogen data from the Rappahannock River.

- Case 3: the concentration of both analytes are reported as less than their reporting limits, that is $c_1 < r_1$ and $c_2 < r_2$. In this case, $I_L = 0$ and $I_U = r_1 + r_2$, which is identical to a left censoring case, where the reporting limit is $r_1 + r_2$.

The estimation of equation 2 now can be accomplished by assigning an interval value of concentration for every observation. Because the estimation is for the logarithm of concentration, for those cases where IL is equal to zero the lower bound on the interval in log space becomes negative infinity. The weights for the estimation are computed exactly as described in Hirsch and others (2010). The method of estimation is an extension of Tobit analysis defined by Tobin (1958) and described more fully in Judge and others (1985). The Tobit analysis requires that the user specify a distributional assumption for the residuals. The Gaussian (normal) distribution was used for all Chesapeake Bay data analyzed in this report. This estimation method returns values for the individual coefficients in equation 2, along with a "scale" parameter, which is the analog of a standard error of the residuals in the usual weighted least squares implementation. The method of handling the retransformation bias differs from that in the previous versions of WRTDS. In those versions, the smearing estimate (Duan, 1983) was used, but that estimate requires computed values for all of the residuals. In the censored case, the residuals are not known for all of the samples. Therefore, the approach was to use the scale to estimate the residual variance and thus

$$\hat{\sigma}_e^2 = scale^2 \qquad (3)$$

Note that in WRTDS (versions 3 and 4) the variance estimate is specific to the particular values of discharge and time for which concentration is being estimated. This is very different from the approach used in ESTIMATOR (Cohn, 2005) in which the variance is assumed constant over all values of discharge and time. This difference in the approach can have substantial effects on the retransformation method and resulting biases. The datasets analyzed in this investigation show wide variations in scale, and these variations typically are very different at high discharges versus low discharges. The retransformation bias correction term α is:

$$\alpha = \exp\left(\frac{\hat{\sigma}_\varepsilon^2}{2}\right) \qquad (4)$$

and the estimate of concentration for each model is obtained by modifying equation 2 to the following:

$$\hat{c} = \alpha \times \exp(\beta_1 + \beta_2 \ln(Q) + \beta_3 t + \beta_4 \sin(2\pi t) + \beta_5 \cos(2\pi t)) \qquad (5)$$

Note that all of the parameters in equation 5 ($\alpha, \beta_1, \beta_2, \beta_3, \beta_4, \beta_5$) are estimates that are the result of each unique weighted survival regression analysis where the weights are based on "distances" between the point of estimation and the sample values of time, discharge, and season. Thus, one important difference between WRTDS and ESTIMATOR is that this retransformation bias correction term in WRTDS can vary by a substantial amount across the range of time, discharge, and season but in ESTIMATOR it is constant.

The survival regression method produces estimates that are virtually identical to those computed by the linear regression and smearing estimator method used in Hirsch and others (2010) even when there is no censoring. For consistency of implementation, this survival regression approach is used, in WRTDS, for all datasets considered, regardless of whether they contained censored data.

Determination of Trends in Water-Quality Conditions

The next point of comparison between the two models is how ESTIMATOR and WRTDS address the question of whether water-quality conditions in the Chesapeake Bay watershed are changing. Two considerations should be addressed when choosing an analytical approach for this purpose. First, natural variation in discharge confounds the determination of water-quality trends. Because water-quality concentrations are highly correlated with discharge, a large portion of the variation in water quality is simply a reflection of the year-to-year variations in discharge. These year-to-year variations in discharge confound the attribution of changes in water quality to land-management practices. ESTIMATOR can generate information about changes in water-quality constituent concentrations that are independent of the random variations in

discharge through a process called "flow adjustment;" while, WRTDS can generate information about changes in both water-quality concentration and flux that are independent of the random variations in discharge through a process called "flow normalization." Both the flow-adjustment and flow-normalization processes are described below in detail.

The second consideration is how water-quality conditions will be represented. Will they be characterized by changes in concentration or by changes in flux or both, and how are the changes calculated? This is important because trends in concentration and flux can tell very different stories and are best suited to different purposes. The computation of an average concentration over a period, such as a year, treats the values for each day equally. The average of 365 daily concentrations is the sum of these values divided by 365. In contrast, the computation of an average (or total) flux over a period, such as a year, gives much more weight to the concentrations on high-discharge days than it does to concentrations on low-discharge days, because the average flux for the year is the sum of the fluxes for each day, which is the product of concentration and discharge for that day, divided by 365. Thus, large changes in concentration on low-discharge days can have a big influence on average concentration but minimal effect on average flux. Conversely, large changes in concentration on high-discharge days can have a big influence on average flux but minimal effect on average concentration. Information on trends in concentration is useful for identifying whether water quality is improving or degrading at a monitoring location and is particularly useful for assessing progress toward attainment of water-quality standards. Information about trends in flux is much more relevant to assessing conditions in a large downstream water body such as the Chesapeake Bay. Trends in flux focus on the total inputs of nutrient and sediment, which can be crucial to ecological conditions in the bay. For example, eutrophication is likely to be most responsive to the total input of nutrients over periods of time such as a year or a season. Accurate information about the trends in concentration and the trends in flux are both vital to assessing the changing influences on the estuary.

The USGS RIM program currently (2012) uses ESTIMATOR to determine trends in flow-adjusted concentration. The flow-adjusted concentration is determined using ESTIMATOR (eq. 1) for a single model window, typically 1985 to present for long-term trends and the most recent 10 years for shorter-term trends. The variability in water-quality concentration that is directly related to flow is removed by the log discharge and log discharge-squared terms. Similarly, variability associated with time of year is removed, using the seasonal terms (eq. 1). The trend in flow-adjusted concentration is determined based on the magnitude, direction, and significance of the time and time-squared coefficients. The flow-adjusted concentration trend is presented as a percentage change that occurred over the entire time period defined by the model window. A full description of the ESTIMATOR flow-adjustment methodology can be found in Langland and others (2006). Note that flow-adjusted concentration and flow-adjusted flux trends presented as a percentage change per year are forced by ESTIMATOR's fixed mathematical form to be identical, which can be an unrealistic model constraint. As a result, changes in water-quality conditions, using ESTIMATOR, are represented using the flow-adjusted concentration only. The CBP has been using flow-adjusted concentration trends, calculated using ESTIMATOR, as an indicator of observed changes in water-quality conditions across the Chesapeake Bay watershed, resulting from human activities. With the recent establishment of the total maximum daily load for the entire Chesapeake Bay watershed, however, the CBP has shifted focus from concentrations to flux of nutrients and sediment to the bay. Accordingly, the CBP has requested that the USGS develop a more appropriate tool to use as an indicator of changing water-quality conditions that more accurately characterizes temporal changes in the flux of nutrients and sediment being delivered to the bay.

WRTDS uses a different approach to characterize temporal changes in water-quality conditions. Instead of a flow-adjustment approach, WRTDS uses "flow normalization" to remove the variability in water-quality conditions that is directly related to the random variations in discharge. The theory behind flow normalization is that the flow that occurred on a given day is one realization from a probability distribution of flows that can be expected for that particular time of year. For computational purposes, that probability distribution is the set of all flows that occurred on that same date throughout the monitoring record, with each considered equally likely to occur. For example, the flow that occurred at the James River RIM station on April 6, 2004, is one realization of 31 possible realizations during the 31-year period 1980 to 2010. WRTDS determines the flow-normalized concentration for April 6, 2004, by estimating 31 possible concentrations by running the model 31 times, using equation 2, each centered in time to April 6, 2004, but with the value of flow equal to each of the 31 observed values. The flow-normalized concentration for April 6, 2004, is the mean of the 31 estimated concentrations. Similarly, WRTDS estimates flow-normalized flux for April 6, 2004, as the mean of the 31 estimated flux values for that date. This process is repeated for every day in the record (31×365) to obtain a daily time series of flow-normalized concentration and flow-normalized flux. Daily flow-normalized flux values are aggregated to obtain monthly and annual total flux. A complete description of the WRTDS flow-normalization method can be found in Hirsch and others (2010).

To summarize, WRTDS produces two complete daily time series of concentration and two daily time series of flux for every modeled water-quality constituent. The first is a time series for both concentration and flux derived directly from the estimation grid of unique time and discharge combinations. The second is a record of both flow-normalized concentration and flux calculated as described in the previous section.

Changes in water-quality conditions, represented in this report as a total annual flux, during any given time period are determined using two different approaches. These two approaches present the change in flux as a slope over a given time period. The first approach is to define the flux change as a slope change per year as follows:

$$Slope \ (Percent) = \left(\left(\left(f_{t_2} - f_{t_1} \right) / f_{t_1} \right) \times 100 \right) / Years \quad (6)$$

where,

f_{t_2} is the total annual flow-normalized flux in year t_2,

f_{t_1} is the total annual flow-normalized flux in year t_1, and

Years   is the total number of years over which the slope is defined ($Years = t_2 - t_1$).

The second approach is to define the flux change as a change in mass (tons per day) per year. To facilitate the comparison of changes in mass between the nine RIM stations, the mass from each watershed was normalized on the basis of the watershed drainage area, thus producing a slope in *yield* with units of tons per day per square mile per year and is calculated as follows:

$$Slope \ (Yield) = \left(\left(\left(f_{t_2} / DA \right) - \left(f_{t_1} / DA \right) \right) / \left(f_{t_1} / DA \right) \right) / Years$$

$$(7)$$

where, *DA* is the watershed drainage area (in square miles) for a given RIM station.

Reporting Error Associated with Flux and Trend Estimates

The last point of comparison between each model's form and function is to evaluate the ability of each model to assign uncertainty to estimates of flux and associated trends in water-quality conditions. The standard error of prediction is the primary statistic used to assign uncertainty to estimates derived from regression models. From the standard error, and statistics used to compute the standard error, the 95-percent confidence interval for flux and trend estimates can be determined. Additionally, using the standard error will enable hypothesis testing to determine if a measured trend in concentration or flux is statistically different from zero (zero indicating no trend). ESTIMATOR does compute a standard error associated with the estimates of flux and flow-adjusted concentration trends. The standard errors are determined following the approach in Gilroy and others (1990) and Cohn and others (1992). As a result, all flux and trend results estimated by ESTIMATOR are presented with associated 95-percent confidence intervals, and the trend results are subjected to hypothesis testing to determine if the trend is statistically different from zero, following the procedure outlined in Langland and others (2006). The reliability of the estimates and associated standard error depends on three assumptions: (1) the model form represents the actual behavior of the system, (2) the error variance of the model is equal across all of the days (regardless of season or discharge), and (3) the errors are uncorrelated in time. Conversely, the estimates of flux and trend in flow-normalized flux, derived from WRTDS, are not associated with a measure of uncertainty. Consequently, all flux and trend estimates derived from WRTDS are not reported within the context of a 95-percent confidence interval; as a consequence, the reported flow-normalized flux trends do not support hypothesis testing to determine if the slope of the trend is significantly different from zero. The lack of an estimate of uncertainty in WRTDS is a function of the complexity of propagating error through each of the 5,838 models created for each water-quality constituent. Research is underway to develop an approach to assign uncertainty to WRTDS-derived estimates.

Comparison of Estimate Accuracy and Bias

The next point of comparison between ESTIMATOR and WRTDS is to identify the accuracy of each model's predictions compared to observations of flux. It should be noted that no model of environmental systems will yield 100 percent accuracy; however, the goal in developing these models is to produce estimates of flux with maximum accuracy and minimal bias. This section evaluates the differences between ESTIMATOR- and WRTDS-derived estimates of nutrient and sediment flux when compared to the sampled data on which they were based. To answer this question, model accuracy and bias associated with ESTIMATOR and WRTDS flux estimates for total nitrogen, nitrate, total phosphorus, orthophosphorus, and suspended sediment at each of the nine RIM stations were compared. Model accuracy is represented by the root mean square error (RMSE) and is defined by the following equation:

$$RMSE = \sqrt{\frac{\sum_{i\,1}^{n} \left(y_i - \hat{y}_i \right)^2}{n-2}} \qquad (8)$$

where

y_i is observed yield, in pounds per day per square mile;

$\hat{y}_i$ is estimated yield, in pounds per day per square mile; and

n is the total number of observations.

As the value of RMSE approaches zero, the model predictions more closely represent actual observations. The values of estimated and observed flux are normalized by drainage area (presented as yield) to facilitate the comparison of RMSE across all nine RIM stations. Model bias (reported here as flux bias) is a measure of the model's tendency to over or underpredict observed fluxes and is defined as:

$$Flux\ Bias = \left(\frac{\sum_{i\,1}^{n} L_{p,i}}{\sum_{i\,1}^{n} L_{o,i}} \right) \qquad (9)$$

where

$L_{p,i}$ is estimated yield for day I, in pounds per day per square mile;

$L_{o,i}$ is observed yield for day i, in pounds per day per square mile; and

n is the total number of observations.

As the value of flux bias approaches 1.0, the model prediction of flux more closely represents the observed flux. Flux bias values greater than 1.0 indicate that the model estimates of flux tend to be greater than observed fluxes, and values less than 1.0 are indicative of the model's tendency to underpredict observed fluxes.

RMSE and flux bias are valuable measures for comparing how well model estimates reflect environmental observations; however, both of these measures are sensitive to the presence of extreme values, as well as the particular pattern of sampled days and would be expected to change if the sampled days included more high-discharge days or more low-discharge days. Therefore, for this investigation, only major differences in RMSE and flux bias are used to signify noteworthy differences between ESTIMATOR and WRTDS. For RMSE, differences between ESTIMATOR and WRTDS are considered significant when there is at least a 20-percent difference (percent difference=(((WRTDS RMSE–ESTIMATOR RMSE)/ESTIMATOR RMSE)×100)). Similarly for flux bias, differences between ESTIMATOR and WRTDS are considered significant when the magnitude of the difference in the two flux-bias ratios is greater than or equal to 0.10. Note that the use of the word "significant" here is not to be interpreted in the statistical sense (that is, a probability of falsely rejecting a null hypothesis). Rather, it is used to refer to an indication of practical significance. Ideally, one would like to test the quality of the two methods using datasets that contain a complete record of daily flux, but such records do not exist within the Chesapeake Bay watershed.

Throughout this section, results are compared for observed versus predicted (either ESTIMATOR or WRTDS) total nitrogen, nitrate, total phosphorus, orthophosphorus, and suspended sediment at each of the nine RIM stations. To simplify the discussion, the term "combinations" refers to the 45 different combinations made up of nine RIM stations and five constituents.

Observed and Estimated Flux Comparison Results

For all 45 possible combinations, RMSE for WRTDS is smaller (more accurate) than the RMSE for ESTIMATOR (table 2). Of the 45 possible combinations, 22 combinations have values of RMSE that show WRTDS is considerably more accurate (20 percent or greater reduction in RMSE) than ESTIMATOR. Of these 22 combinations, 7 show

Table 2. Measures of RMSE and flux bias relating WRTDS- and ESTIMATOR-derived nutrient and sediment fluxes compared to discrete flux observations at the nine River Input Monitoring (RIM) stations

[(lbs/day)/mi^2, pounds per day per square mile; **bold text**, percent difference between 20 and 40; **bold text**, percent difference between 41 and 60; **bold text**, percent difference greater than 61; text, WRTDS flux bias ratio closer to 1.0 by at least 0.10; **bold text**, difference between WRTDS and ESTIMATOR flux bias ratio 0.10 or greater; Category, variable used to group station and constituent combinations based on differences between WRTDS and ESTIMATOR flux-bias results]

RIM station	Root mean squared error ((lbs/day)/mi^2)			Flux bias ratio			Category
	WRTDS	ESTIMATOR	Percent difference	WRTDS	ESTIMATOR	Difference	
Total nitrogen							
Susquehanna	25.22	64.44	**−61**	0.99	0.95	0.04	I
Potomac	29.06	38.42	**−24**	0.99	0.95	0.04	I
James	12.31	12.94	−5	1.01	1.00	0.01	I
Rappahannock	12.55	21.50	**−42**	1.00	1.12	**−0.12**	II
Appomattox	2.32	2.57	−10	1.01	1.02	−0.01	I
Pamunkey	3.59	4.27	−16	0.98	1.03	−0.05	I
Mattaponi	1.62	1.70	−4	1.00	1.03	−0.03	I
Patuxent	7.02	7.32	−4	1.00	1.01	−0.01	I
Choptank	11.85	13.31	−11	0.99	1.02	−0.03	I
Nitrate							
Susquehanna	6.41	6.65	−4	0.99	1.00	−0.01	I
Potomac	5.81	7.01	−17	0.98	1.03	−0.05	I
James	1.46	2.03	**−28**	0.99	1.12	**−0.13**	II
Rappahannock	2.94	6.70	**−56**	0.99	1.36	**−0.37**	II
Appomattox	0.89	1.16	**−23**	1.03	1.09	−0.06	I
Pamunkey	0.90	1.12	**−20**	0.98	1.04	−0.06	I
Mattaponi	0.61	0.70	−13	1.00	1.02	−0.02	I
Patuxent	7.02	7.32	−4	1.00	1.01	−0.01	I
Choptank	8.01	8.52	−6	0.97	1.01	−0.04	I
Total phosphorus							
Susquehanna	1.57	2.82	**−44**	1.06	0.96	**0.10**	III
Potomac	8.54	11.41	**−25**	1.03	1.12	−0.09	I
James	4.91	7.90	**−38**	1.03	1.07	−0.04	I
Rappahannock	6.36	23.88	**−73**	1.02	1.53	**−0.51**	II
Appomattox	0.43	0.45	−4	1.02	1.03	−0.01	I
Pamunkey	0.79	0.95	−17	0.98	1.03	−0.05	I
Mattaponi	0.25	0.27	−7	1.00	1.03	−0.03	I
Patuxent	1.93	2.41	**−20**	1.04	1.01	0.03	I
Choptank	1.42	4.07	**−65**	1.04	1.17	**−0.13**	II
Orthophosphorus							
Susquehanna	0.20	0.22	−9	1.05	1.16	**−0.11**	II
Potomac	0.45	1.21	**−63**	1.02	1.34	**−0.32**	II
James	0.39	0.40	−3	0.99	1.01	−0.02	I
Rappahannock	0.25	0.50	**−50**	1.02	1.14	**−0.12**	II
Appomattox	0.08	0.09	−11	1.03	1.04	−0.01	I
Pamunkey	0.13	0.14	−7	1.00	1.00	0.00	I
Mattaponi	0.06	0.07	−1	0.98	1.01	−0.03	I
Patuxent	0.23	0.25	−8	1.09	1.04	0.05	I
Choptank	0.44	1.03	**−57**	1.06	1.15	−0.09	I

Table 2. Measures of RMSE and flux bias relating WRTDS- and ESTIMATOR-derived nutrient and sediment fluxes compared to discrete flux observations at the nine River Input Monitoring (RIM) stations.—Continued

[(lbs/day)/mi², pounds per day per square mile; **bold text**, percent difference between 20 and 40; **bold text**, percent difference between 41 and 60; **bold text**, percent difference greater than 61; text, WRTDS flux bias ratio closer to 1.0 by at least 0.10; **bold text**, difference between WRTDS and ESTIMATOR flux bias ratio 0.10 or greater; Category, variable used to group station and constituent combinations based on differences between WRTDS and ESTIMATOR flux-bias results]

RIM station	Root mean squared error ((lbs/day)/mi²)			Flux bias ratio			Category
	WRTDS	ESTIMATOR	Percent difference	WRTDS	ESTIMATOR	Difference	
Suspended sediment							
Susquehanna	2,758	9,632	**−71**	1.06	0.80	**0.26**	III
Potomac	6,883	26,536	**−74**	1.15	1.79	**−0.64**	II
James	3,271	12,153	**−73**	1.21	1.80	**−0.59**	II
Rappahannock	8,632	41,288	**−79**	1.32	2.57	**−1.25**	II
Appomattox	145	174	−17	1.02	1.13	**−0.11**	II
Pamunkey	1,229	1,370	−10	0.98	1.07	−0.09	I
Mattaponi	155	158	−2	0.94	1.13	**−0.19**	II
Patuxent	1,396	2,379	**−41**	0.99	1.17	**−0.18**	II
Choptank	609	1,122	**−46**	1.02	1.21	**−0.19**	II

a 20 to 40 percent reduction, 7 show a 41 to 60 percent reduction, and 8 show a 61 percent or greater reduction/improvement in model accuracy when WRTDS is used to estimate flux. The remaining 23 combinations show that the RMSE for WRTDS is marginally smaller (0 to 19 percent) when compared to the RMSE for ESTIMATOR. These RMSE results indicate that WRTDS generates flux estimates that are universally more accurate than flux estimates generated by ESTIMATOR and in nearly half of the combinations the improvement in model accuracy is considerable. The RMSE also can be viewed here as a surrogate for uncertainty in the model estimates where WRTDS would show a reduction in estimate uncertainty (that is, reduction in the width of the confidence interval) compared to ESTIMATOR predictions.

For 36 of the possible 45 (80 percent) combinations, WRTDS generated fluxes that were less biased (that is, flux-bias ratio closer to 1.0) than fluxes generated by ESTIMATOR (table 2). Of these 36 combinations, there were 16 combinations where WRTDS exhibited significantly less bias than ESTIMATOR (closer to 1.0 by at least 0.1). There were 7 of the 45 (16 percent) combinations where ESTIMATOR-generated fluxes were less biased than fluxes generated by WRTDS; however, the biases associated with ESTIMATOR and WRTDS-derived fluxes, for these 7 combinations, were both within 0.10 of 1.0. The remaining 2 of the 45 combinations, WRTDS and ESTIMATOR fluxes were equally biased and within 0.10 of 1.0. These results show that for the majority of the combinations (67 percent, 30 of 45 combinations) the differences in flux bias from 1.0 were marginal. For 33 percent of the combinations

(15 of 45 combinations), however, WRTDS-derived fluxes showed a marked improvement in flux bias (flux bias closer to 1.0) compared to fluxes generated using ESTIMATOR.

Sources of Flux Bias Discrepancies

The results of the flux bias analyses provide valuable information on the overall tendency of WRTDS and ESTIMATOR to over or underpredict fluxes. To better understand how to interpret these flux bias results, however, the 45 combinations were categorized by presumed source of discrepancy between WRTDS and ESTIMATOR. The first two categories are established on the basis of the comparison of flux bias between WRTDS and ESTIMATOR (that is, the total difference between WRTDS and ESTIMATOR flux biases) (table 2) and are (1) marginal difference, less than 0.10 (Category I) and (2) considerable improvements (greater than or equal to 0.10) of WRTDS over ESTIMATOR flux-bias ratio (Category II). The third category contains two combinations of improvements of flux-bias ratio (by more than 0.10) of WRTDS over ESTIMATOR; however, for both of these combinations, ESTIMATOR has a tendency to underpredict flux and WRTDS tends to overpredict flux (Category III). Within each category, the differences between ESTIMATOR and WRTDS predictions of flux are investigated. In addition to flux bias, patterns in the concentration versus discharge (CQ) plots and concentration residual versus discharge (residual) plots were used to illustrate model fit. CQ and residual plots for each RIM combination are provided in appendix 1.

Category I Combinations

Category I combinations are those where the analysis of model bias yielded only marginal differences (that is, less than 0.10 in absolute magnitude) between WRTDS and ESTIMATOR-derived fluxes compared to discrete observations. Twenty-eight of 45 RIM combinations exhibit Category I type results (table 2). A representative combination for all Category I combinations is nitrate at Patuxent River near Bowie, Maryland. This combination was chosen because of the complex concentration-discharge relation exhibited for nitrate at this location (fig. 3A). The Patuxent River is a point-source dominated system receiving discharge of effluent from multiple sewage treatment plants (Sprague and others, 2000). The concentration versus discharge relation for nitrate has been altered as a result of various sewage treatment plant upgrades that specifically reduce nitrate from effluent (for example, biological nitrate reduction (BNR)). The implications of these management actions are evident based on the reduction in nitrate concentration associated with low discharges that range between 50 and 400 ft³/s (cubic feet per second; between 4 and 6 log units; fig. 3A). Prior to these management actions (1981–89), the concentration of nitrate in the Patuxent River ranged from approximately 2.0 to 6.0 mg/L (approximately 0.7 to 2.0, log units) at low discharges; this is evident as the upper left lobe of the concentration-discharge relation in figure 3(A). Following the most recent upgrade of sewage treatment plants to BNR during the early 1990s, nitrate concentrations associated with low discharges range from approximately 1.0 to 2.0 mg/L (approximately 0.0 to 0.70, log units); this is evident as the lower left lobe of the concentration-discharge relation in figure 3(A). Figure 3B shows this same observed concentration-discharge relation for nitrate in the Patuxent (red dots) with the ESTIMATOR-predicted concentrations overlain (black dots). Figure 3D shows the observed concentration-discharge relation (red dots) overlain by WRTDS-predicted concentrations (black dots). Both of these plots show that ESTIMATOR and WRTDS accurately reproduce the complex nature of the concentration-discharge relation for nitrate at the Patuxent. The residual (observed minus predicted concentration) plots for ESTIMATOR and WRTDS (figures 3C and 3E) illustrate changes in model error as a function of changing discharge. The desired pattern in residual plots is a homoscedastic distribution around zero across the full range of discharge. The residual plots shown in figure 3C and 3E indicate that ESTIMATOR and WRTDS, respectively, produced estimates of nitrate concentrations that were similar in accuracy. This example for nitrate at the Patuxent RIM station is representative of all Category I combinations in that both ESTIMATOR and WRTDS produce estimates of concentration and flux that are similar and the differences between flux-bias ratios are marginal.

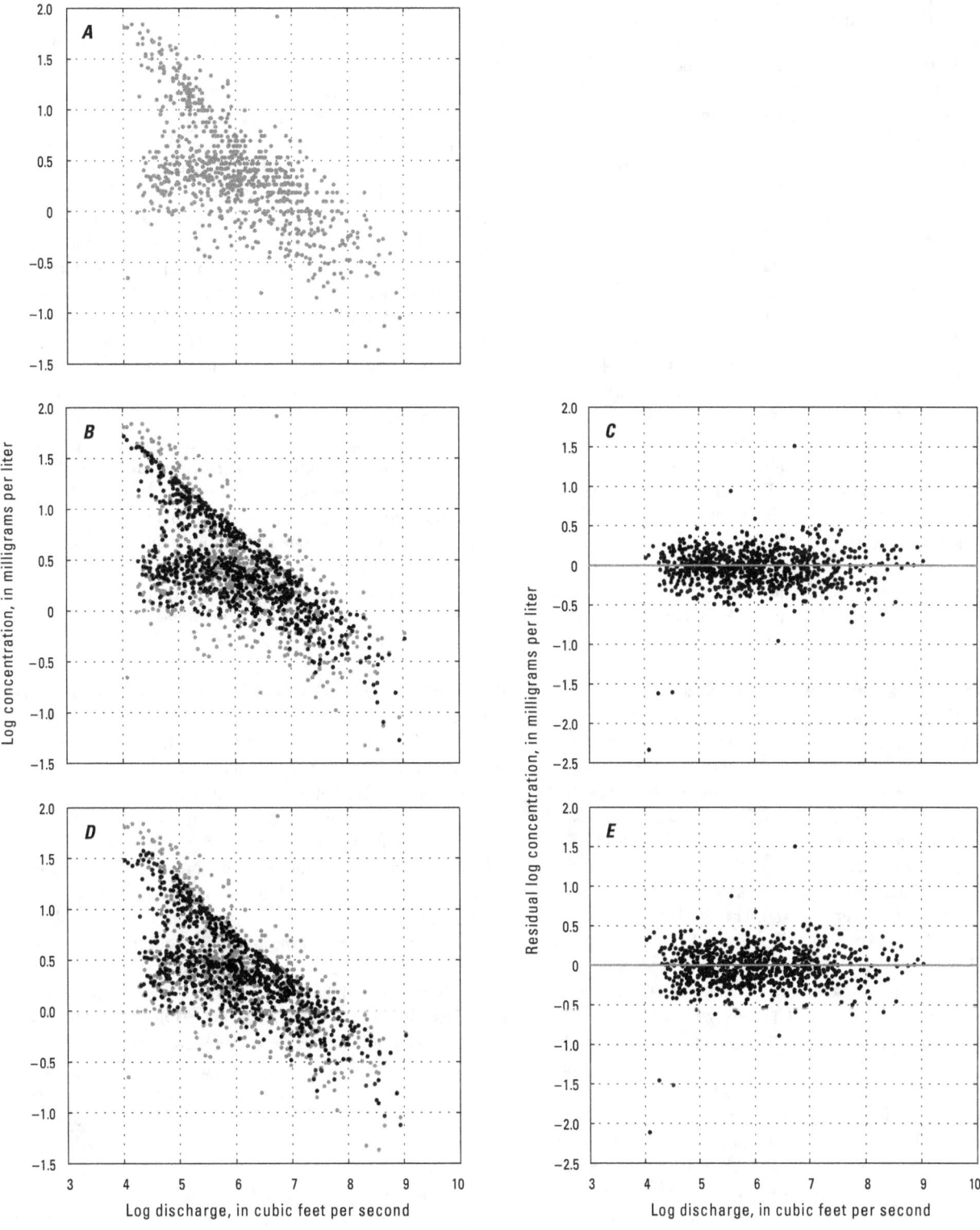

Figure 3. Nitrate at Patuxent River near Bowie, Maryland (USGS Station 01594440), showing the *(A)* observed concentration (red dots) versus discharge relation, *(B)* observed (red dots) and ESTIMATOR-predicted (black dots) concentration versus discharge relation, *(C)* residual (observed minus predicted) plot for ESTIMATOR predictions, *(D)* observed (red dots) and WRTDS-predicted (black dots) concentration versus discharge relation, and *(E)* residual (observed minus predicted) plot for WRTDS predictions.

Category II Combinations

Category II combinations are those where WRTDS-derived estimates of flux are significantly less biased compared to ESTIMATOR-derived fluxes. A total of 15 RIM combinations exhibit Category II type discrepancies (table 2). Two examples illustrate situations that lead to this type of discrepancy and provide insight for potential causes. The first situation, which occurs in 13 of 15 cases of Category II, is when ESTIMATOR considerably overpredicts concentrations associated with high-discharge conditions. A representative example is suspended sediment at the Rappahannock River RIM station. The concentration-discharge relation (fig. 4A) shows sediment concentration (red dots) increases as discharge increases; however, the slope of the relation between log of sediment concentration and the log of discharge decreases when the log of discharge is above about 9 (equivalent to about 8,000 ft^3/s). ESTIMATOR reasonably reproduces the concentration-discharge relation for discharge conditions less than about 8,000 ft^3/s; however, ESTIMATOR considerably overestimates sediment concentration for discharge conditions greater than about 8,000 ft^3/s (fig. 4B). The residual plot shown in figure 4C also clearly shows the overprediction (negative residuals) of suspended-sediment concentration for discharge values greater than about 9 log units. A discharge value of about 8,000 ft^3/s occurs frequently during storm-runoff periods. For context, a discharge value of 25,000 ft^3/s (10.1 log units) at the Rappahannock River station has a 2-year flood-recurrence interval. This peak observed sediment concentration of 1,204 mg/L was predicted by ESTIMATOR to be 4,105 mg/L. This overprediction of sediment concentration, during high-discharge conditions, causes an overprediction of flux, which explains the high flux-bias ratio of 2.57 (table 2). Conversely, suspended-sediment concentrations, estimated by WRTDS, more accurately reflect the shape of the observed concentration-discharge relation throughout the full range of flows from the Rappahannock River (fig. 4D). The WRTDS prediction of the highest observed concentration was 1,423 mg/L. Although WRTDS has a tendency also to overpredict suspended-sediment concentration at discharge values greater than about 8,000 ft^3/s (fig. 4E), which is the primary reason for the flux-bias ratio of 1.32 (table 2), this overprediction of flux by WRTDS is considerably less than ESTIMATOR flux estimates during these same high-discharge conditions.

The second situation of Category II combination occurs when ESTIMATOR underpredicts concentrations that occur during extreme low-discharge and high-discharge periods and overpredicts concentrations during data rich intermediate-discharge conditions. Of the 15 Category II combinations, only 2 combinations exhibit this type of discrepancy; these 2 combinations are nitrate at the James River RIM station and nitrate at the Rappahannock River RIM station. Nitrate concentration at the Rappahannock River RIM station will be used as an example to define this situation and to discuss the root causes. The observed concentration-discharge relation for nitrate at the Rappahannock River RIM station

(fig. 5A) has three distinguishing features. The first and most important feature is that over a very broad range of discharge values, from about 400 to 22,000 ft^3/s (about 6 to 10 in log units), nitrate concentrations appear to be essentially unrelated to discharge and cluster in a range between about 0.2 and 1 mg/L (about −1.5 to 0 log units). The second feature is that during low discharges that range from about 7 to 400 ft^3/s (2 to 6 log units), nitrate concentrations increase with discharge from about the reporting limit of 0.004 mg/L (−6.2 log units) to about 0.2 mg/L (−1.5 log units). These very low concentrations are probably due to some combination of the following factors: (1) the water is derived from deeper groundwater, which may not be subjected to anthropogenic increases in nitrate to the same extent as shallow groundwater, and has had more opportunity for subsurface denitrification on its way to the stream; (2) the river has a low ratio of volume to streambed surface area providing more opportunity for denitrification, which typically occurs at the interface between the riverbed and the water column (hyporheic zone); and (3) typically at times of high temperature, nitrate uptake by aquatic biota may greatly reduce the amount of nitrate remaining in the river. The third feature is that at discharge values above about 20,000 ft^3/s (about 10 log units), the slope of the relation between log concentration and log discharge is negative. This negative slope most likely arises from one or both of two causes: (1) at these high discharges, the water in the river reflects the chemistry of the rain water, rather than the more nitrate-rich soil water and groundwater that dominate streamflow in the lower to middle range of discharges; and (2) the steeper, more forested portions of the watershed become more dominant contributors to runoff at these high discharges and thus waters tend to be lower in nitrate than the waters derived from the more gently sloping parts of the landscape where urban and agricultural activities are more dominant. Because ESTIMATOR is constrained to use the quadratic function to fit this relation, the data have very few observations in this high discharge range, and the anomalous observations cannot be accounted for by time or season. Thus the fit in this range is largely based on the fit in the middle to low discharge range. As shown in figure 5B, ESTIMATOR overpredicts nitrate concentrations for discharges between 1,000 ft^3/s (7 log units) and 22,000 ft^3/s (10 log units) and underpredicts concentrations for discharges above about 30,000 ft^3/s (about 10.3 log units). The residual plot shown in figure 5C also clearly shows that ESTIMATOR underpredicts (positive residuals) nitrate at low and high discharges and overpredicts nitrate concentrations during intermediate discharges. This overprediction of nitrate concentration for discharges between about 1,000 and 22,000 ft^3/s (about 7 to 10 log units) directly contributes to the flux-bias ratio of 1.36 (table 2). WRTDS produces nitrate concentrations that more closely follow the patterns exhibited in the observed nitrate concentration-discharge relation (fig. 5D), because it is not constrained to follow the quadratic functional form. Residuals are produced that are much more symmetrical around zero over most of

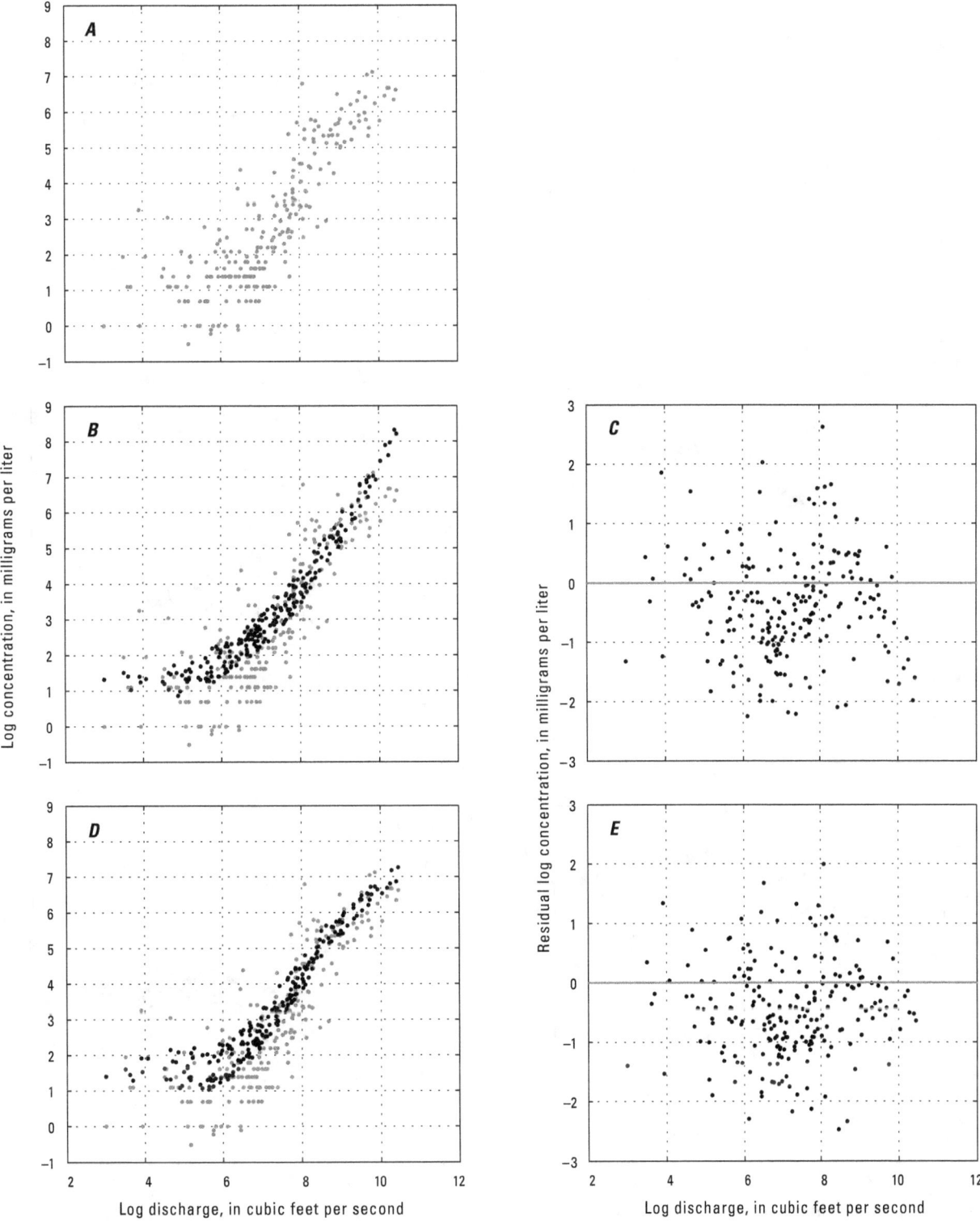

Figure 4.   Suspended-sediment concentration at Rappahannock River near Fredericksburg, Virginia (USGS Station ID 01668000), showing the *(A)* observed concentration (red dots) versus discharge relation, *(B)* observed (red dots) and ESTIMATOR-predicted (black dots) concentration versus discharge relation, *(C)* residual (observed minus predicted) plot for ESTIMATOR predictions, *(D)* observed (red dots) and WRTDS-predicted (black dots) concentration versus discharge relation, and *(E)* residual (observed minus predicted) plot for WRTDS predictions.

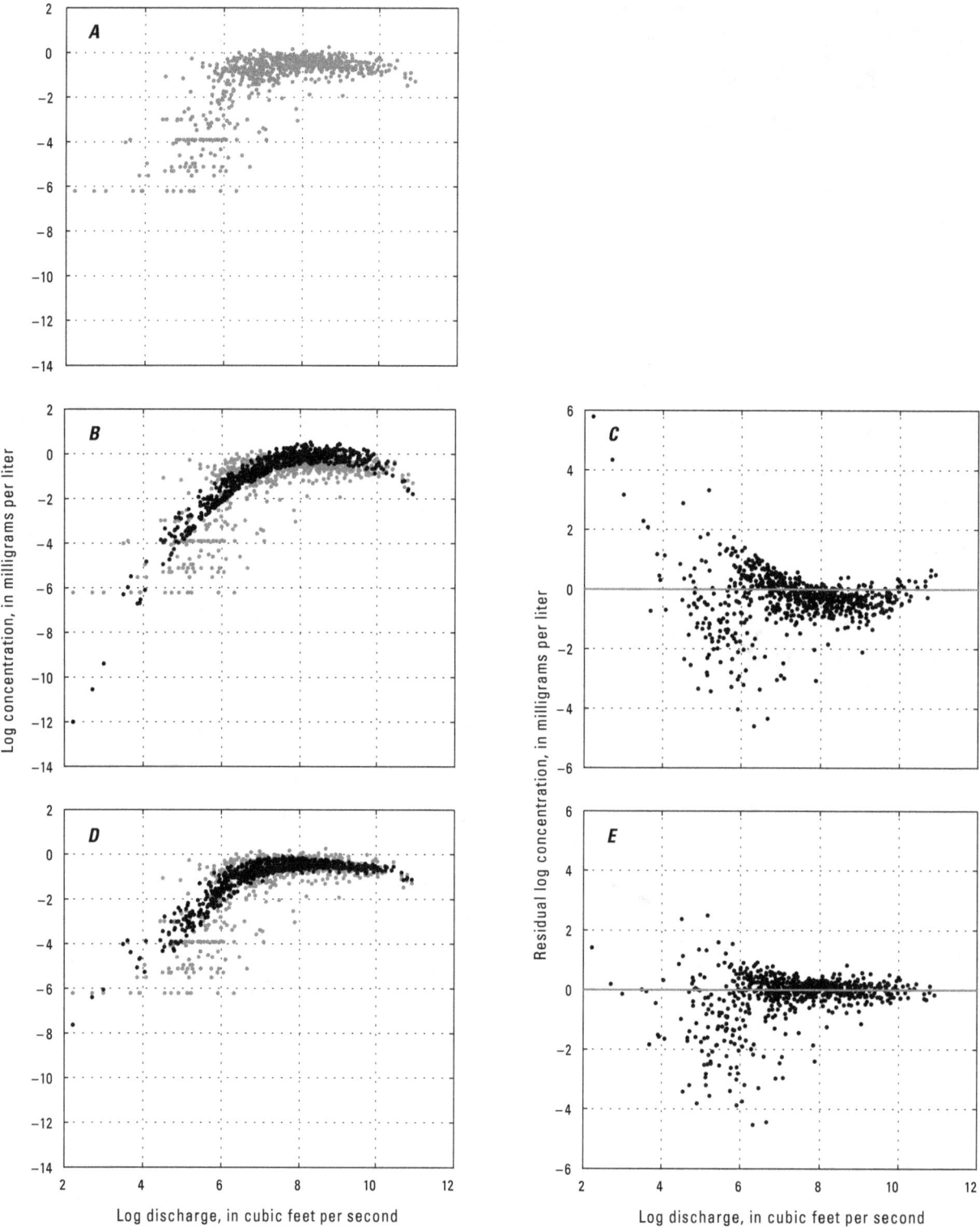

Figure 5.   Nitrate concentration at Rappahannock River near Fredericksburg, Virginia (USGS Station ID 01668000), showing the *(A)* observed concentration (red dots) versus discharge relation, *(B)* observed (red dots) and ESTIMATOR-predicted (black dots) concentration versus discharge relation, *(C)* residual (observed minus predicted) plot for ESTIMATOR predictions, *(D)* observed (red dots) and WRTDS-predicted (black dots) concentration versus discharge relation, and *(E)* residual (observed minus predicted) plot for WRTDS predictions.

the range of discharges (certainly those above about 7 log units or 1,100 ft³/s). The residuals for both methods appear to be highly heteroscedastic. In either method the residual variance is fairly small and constant above a discharge of about 1,000 ft³/s (about 7 log units), but it is much larger for the lower discharge values (less than 7 log units) (fig. 5C and 5E). Note that ESTIMATOR assumes that the residuals have constant variance and uses that overall variance in making the bias adjustment (see Cohn, 2005). This assumption is likely to result in too large a bias adjustment at high discharges and too small a bias adjustment for concentrations at low discharges. In addition to the general problem of lack of fit, this assumption also contributes to the positive flux bias associated with ESTIMATOR. In contrast, WRTDS makes no assumption of constant variance, and the bias adjustment (eq. 4) tends to be small at high discharges and large at low discharges. As a result of the improved accuracy associated with the WRTDS nitrate predictions, the flux-bias ratio improved from 1.36 (ESTIMATOR predictions) to 0.99 (table 2).

Both ESTIMATOR (fig. 5C) and WRTDS (fig. 5E) show relatively poor ability to predict log nitrate concentrations at low discharge values. Although the graphs in figure 5 indicate a high degree of variability at low discharge, it should be noted that these are graphs of log concentration. Concentrations at these low discharges are generally between 0 and 0.1 mg/L and most concentrations at middle to high discharges range from 0.2 to 1.0 mg/L. This relatively poor fit (using either method) is of little consequence in the calculation of annual or long-term average fluxes, because the concentrations are so close to zero that they contribute little to the total flux. What is important from the standpoint of the performance of the two methods is that the inability of both methods to accurately predict concentrations in this low discharge range influences the predictions by ESTIMATOR at the much more important high discharges but does not affect predictions by WRTDS at high discharges.

Both examples of Category II discrepancies highlight that although ESTIMATOR does an exceptional job accurately estimating concentration-discharge relations that are linear or quadratic in nature (Category I combinations), ESTIMATOR is unable to fully reproduce concentration-discharge relations that have multiple points of inflection (that is, more sinuous in form). The primary reason that ESTIMATOR cannot fully reproduce concentration-discharge relations for Category II combinations is because these relations are not linear or quadratic, and the deviation in the relation cannot be accounted for by the remaining model variables, time and season. WRTDS, conversely, more closely reproduces the observed shape of the concentration-discharge relation because WRTDS is not constrained to linear or quadratic forms, but rather uses smoothing techniques to more closely approximate the sinuous shape of the relation. Additionally, data from low discharge days have limited influence on the model coefficients for days of high discharge because the weights assigned to these observed data collected during low-discharge conditions are at or near zero (and the reverse

also is true). This is not the case with ESTIMATOR where all water-quality data are used to determine model coefficients for each 9-year model window.

Category III Combinations

Category III contains two combinations, Susquehanna River total phosphorus and Susquehanna River suspended sediment, that exhibit some features consistent with the sources of flux bias discrepancies discussed in Category II; however, other unique characteristics warrant labeling these two combinations as Category III. The general characteristics that define Category III combinations are (1) ESTIMATOR underpredicts flux associated with the highest discharges; (2) WRTDS overpredicts flux associated with moderate discharges; (3) the magnitude of the difference between flux-bias ratio associated with WRTDS and the flux-bias ratio associated with ESTIMATOR is at least 0.10. The net effect of these three characteristics is that WRTDS results will show that larger fluxes of suspended sediment and total phosphorus are being delivered from the Susquehanna River RIM station when compared to ESTIMATOR.

Susquehanna River suspended-sediment concentrations will be used to represent both Category III combinations; the patterns and discrepancies associated with total phosphorus are similar but less pronounced as those associated with Susquehanna River suspended sediment. The flux-bias ratio for Susquehanna River suspended sediment increases from 0.80 to 1.06 (net change of 0.26) when ESTIMATOR flux predictions are replaced by those derived from WRTDS (table 2). This flux-bias ratio indicates that WRTDS produces flux estimates that are positively biased (1.06) but are closer to 1.0 than the flux-bias ratio associated with ESTIMATOR, which is negatively biased (0.80). To better understand the reasons associated with the ESTIMATOR and WRTDS-associated flux biases, one needs to take a close look at the observed relation between log of suspended-sediment concentration and log of discharge (fig. 6A). The concentration-discharge relation (fig. 6A) shows that suspended-sediment concentration (red dots) increases as discharge increases; however, the slope of the relation between log of suspended-sediment concentration and the log of discharge steepens at two separate locations. The first increased slope is represented for log of discharges between 11 and 13 (equivalent to discharges between about 60,000 and 450,000 ft³/s); and an additional increase occurs at log of discharges greater than about 13 (albeit only four observations are defining this increased slope). The Susquehanna River RIM station is located at the outfall of Conowingo reservoir. The changing shape of the concentration-discharge relation is related to an abrupt shift from net depositional at lower discharges to net scour of the reservoir sediments at higher discharges. Langland (2009) estimated that a minimum discharge threshold of at least 390,000 ft³/s must be exceeded to mobilize the stored reservoir sediments. ESTIMATOR accurately reproduces the concentration-discharge pattern for all discharge values up

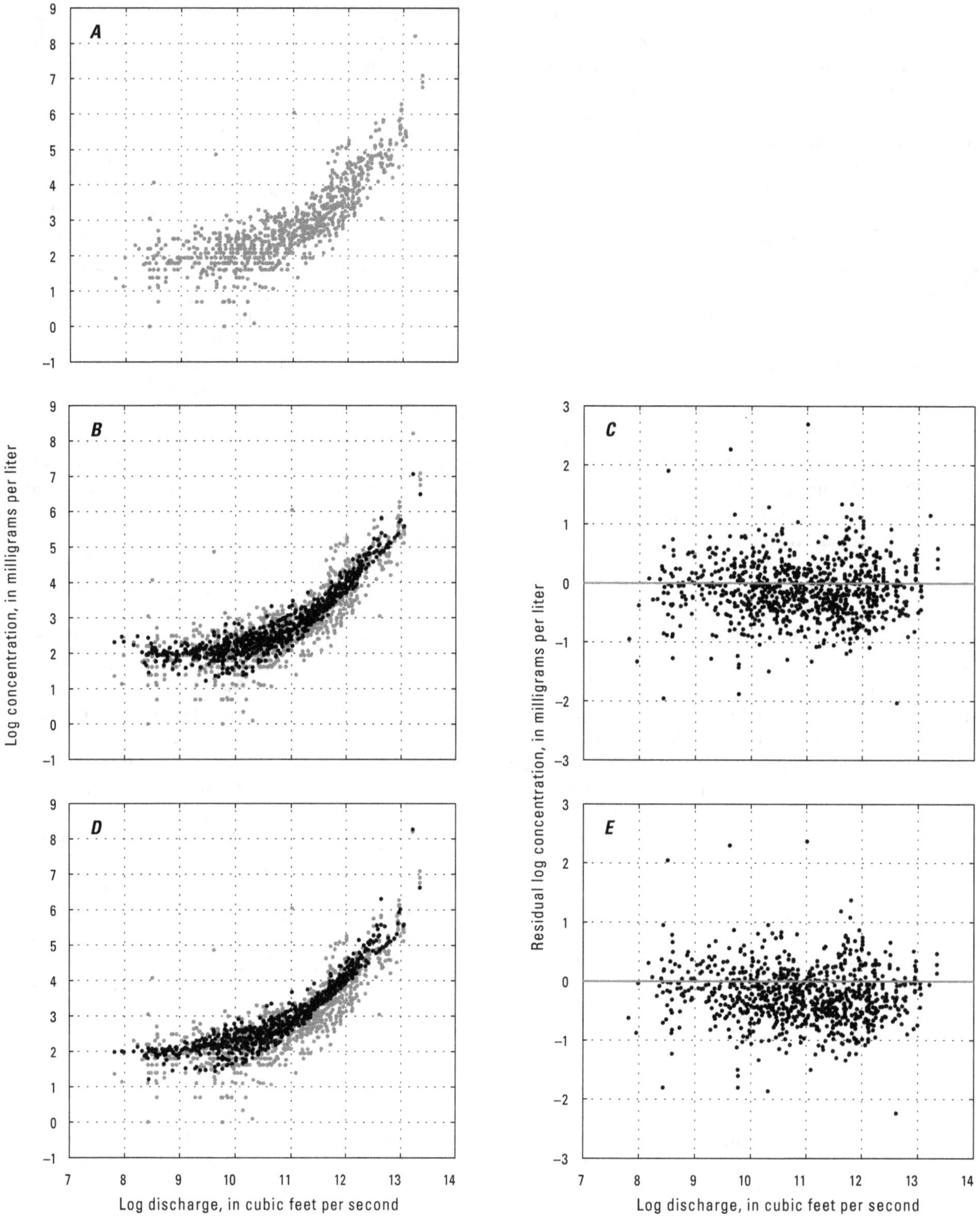

Figure 6.   Suspended-sediment concentration at Susquehanna River near Conowingo, Maryland (USGS Station ID 01578310), showing the *(A)* observed concentration (red dots) versus discharge relation, *(B)* observed (red dots) and ESTIMATOR-predicted (black dots) concentration versus discharge relation, *(C)* residual (observed minus predicted) plot for ESTIMATOR predictions, *(D)* observed (red dots) and WRTDS-predicted (black dots) concentration versus discharge relation, and *(E)* residual (observed minus predicted) plot for WRTDS predictions.

to about 450,000 ft³/s (13 log units); however, ESTIMATOR tends to underestimate suspended-sediment flux for discharges greater than about 450,000 ft³/s (fig. 6B). It is this underprediction of suspended-sediment concentration associated with the highest discharges (that is, greater than 450,000 ft³/s) that primarily produces the flux-bias ratio of 0.80 (table 2). The residuals for the ESTIMATOR predictions are evenly distributed around zero for discharges up to 450,000 ft³/s (fig. 6C); conversely; the residuals skew to greater than zero (indicative of underestimation of flux by ESTIMATOR) for discharges greater than 450,000 ft³/s. The primary reason that ESTIMATOR underpredicts suspended-sediment concentrations associated with discharges greater than 450,000 ft³/s is that the observed concentration-discharge relation is not truly quadratic in form at high discharges (greater than approximately 400,000 ft³/s, or 13 log units). WRTDS is more successful than ESTIMATOR at reproducing the concentration-discharge relation for discharges greater than about 450,000 ft³/s; however, WRTDS is not as successful as ESTIMATOR at reproducing log suspended-sediment concentrations associated with log discharges between approximately 11 and 12.5 (equivalent to discharges between about 60,000 and 270,000 ft³/s) (fig. 6D). It is this overprediction of suspended-sediment concentration for discharges between 60,000 and 270,000 ft³/s that produces the flux-bias ratio of 1.06 (table 2). The reason for the overprediction of suspended-sediment concentration for discharges between 60,000 and 270,000 ft³/s by WRTDS may be related to lack of a log discharge squared term.

What can be concluded from this comparison of WRTDS and ESTIMATOR model estimates of flux to the fluxes observed at the nine RIM stations? For 28 of the 45 RIM combinations (Category I), there is little to no difference between the ESTIMATOR and WRTDS estimates of flux with regard to flux-bias ratio; however, for all 28 combinations, WRTDS produced flux estimates that were more accurate (reduced RMSE) than those derived from ESTIMATOR with 8 of the 28 combinations having greater than 20 percent reduction in RMSE. For 15 of the 45 combinations, WRTDS fluxes were less biased (flux-bias ratio closer to 1.0 by at least 0.10) and had a marked improvement in accuracy (12 of the 15 combinations showed a 20-percent reduction in RMSE) compared to flux estimates obtained from ESTIMATOR (Category II). These results show that ESTIMATOR's strength is reproducing concentration-discharge relations (in log space) that are either linear or quadratic in form; the majority (62 percent or 28 of 45 combinations) of the combinations exhibits this relation. The problem arises, for ESTIMATOR, when the concentration-discharge relation is more sinuous (more than one point of inflection) in form than a simple linear or quadratic pattern, and the differences between the observed concentration-discharge relation and a linear or quadratic form cannot be accounted for by the additional model variables, time and season. WRTDS, conversely, is able to represent these complex concentration-discharge relations because of the model flexibility brought about by the combination of its functional form and how the model coefficients are determined for every unique combination of time and discharge.

Only 2 combinations of the 45 cannot be categorized as Category I or Category II. What makes these Category III combinations unique is that they are either particulate or particulate-dominated constituents (suspended sediment and total phosphorus) collected from the Susquehanna River downstream from the Conowingo reservoir. ESTIMATOR tends to outperform WRTDS in reproducing the concentration and flux associated with the majority of the range in discharge; WRTDS outperforms ESTIMATOR in reproducing concentrations and fluxes associated with the highest discharges. The difficulty that both models have with reproducing concentrations and fluxes across the full range of discharges is directly related to transport processes that change (shift from net depositional to net scour) depending on conditions in the Conowingo reservoir. In this situation, WRTDS seems to be the better model to address questions associated with flux and changes in flux, and ESTIMATOR seems to be better at reproducing fluxes and concentrations associated with non-extreme hydrologic events.

Comparison of ESTIMATOR and WRTDS Annual Fluxes

The results of the analysis thus far have shown that flux estimates derived from WRTDS are in all cases except one more accurate and for the majority of RIM combinations have an improved flux-bias ratio when compared to flux estimates derived from ESTIMATOR. The focus of the analysis shifts now to address how different estimates of annual fluxes would have been for all RIM combinations had WRTDS been used instead of ESTIMATOR for estimates for 1980 to 2010. To answer this question the average percent difference in annual flux estimates was quantified between WRTDS-derived and ESTIMATOR-derived annual fluxes of total nitrogen, nitrate, total phosphorus, orthophosphorus, and suspended sediment at each of the nine RIM stations. The average difference in estimated annual fluxes (expressed in percent) is defined as

$$\text{Average Difference} = \left(\frac{\sum_{i\,1}^{n} AF_{W,i} - \sum_{i\,1}^{n} AF_{E,i}}{\sum_{i\,1}^{n} AF_{E,i}} \right) \times 100 \tag{10}$$

where

$AF_{W,i}$ is WRTDS estimated annual flux for year i; in tons per day;

$AF_{E,i}$ is ESTIMATOR estimated annual flux for year i, in tons per day; and

n is the total number of years.

Positive values for the average percent difference in annual flux indicate that annual fluxes would have been, on average, greater coming from WRTDS; and, negative values indicate that, on average, annual fluxes derived from WRTDS would have been smaller than those derived from ESTIMATOR. An average percent difference less than 10 percent is considered minimal.

Nitrogen

Total nitrogen annual fluxes generated by WRTDS for the nine RIM stations are consistent with those generated using ESTIMATOR (fig. 7). All stations show a close correspondence for annual fluxes of total nitrogen derived from WRTDS and ESTIMATOR. Average percent differences for annual fluxes of total nitrogen range from –5.13 to 3.04 percent (fig. 7). These differences are all less than 10 percent and are considered minimal. The greatest average difference (–5.13 percent) occurs at the Rappahannock RIM station (fig. 7D). The Rappahannock is the only station that exhibits Category II discrepancies that are related to overprediction of flux, by ESTIMATOR, during the highest discharge conditions. Recall that flux-bias ratios for total nitrogen at the Rappahannock RIM station are 1.00 (no bias) and 1.12

for WRTDS and ESTIMATOR-generated fluxes, respectively (table 2). Figure 7D shows that the greatest discrepancy between ESTIMATOR and WRTDS annual fluxes occurs during years with the greatest flux. The highest total nitrogen annual flux, between 1985 and 2010, occurred during 2003 when the annual flux determined by ESTIMATOR was 15.5 tons per day (tons/d) and the annual flux determined by WRTDS was 13.6 tons/d (fig. 7D). The annual discharge for 2003 is the greatest annual discharge for this same period. Thus, the effect of ESTIMATOR's tendency to overpredict concentrations associated with the highest discharges becomes most apparent during the years with the greatest discharges. The average difference for total nitrogen at the Rappahannock RIM station is interpreted as follows: historical annual fluxes would have been, on average, 5.13 percent lower coming from WRTDS compared to annual fluxes generated using ESTIMATOR, and much of this difference comes from the estimates in the years for which ESTIMATOR produced its two highest annual values of the entire period of record.

Nitrate annual fluxes generated by WRTDS for the nine RIM stations are consistently less than those generated using ESTIMATOR (fig. 8). Average differences for annual fluxes of nitrate range from –28.13 to –1.61 percent (fig. 8). The average differences are all negative; this means the annual nitrate fluxes derived from WRTDS are typically smaller than those derived from ESTIMATOR. The two stations with the greatest average percent difference are the Rappahannock (–28.13 percent) (fig. 8D) and James (–13.55 percent) (fig. 8C). Both of these stations, being identified as having Category II discrepancies for the flux-bias ratio, have large negative average differences because ESTIMATOR overpredicts nitrate flux during intermediate- to high-discharge conditions, and WRTDS-derived fluxes more accurately represent observed nitrate concentrations/fluxes for this discharge interval. The remaining seven RIM stations, listed as having Category I-type discrepancies, have average percent differences that range from –8.38 to –1.61 percent, which are considered to be minimal differences because they are all less than 10 percent different.

Phosphorus

Total phosphorus annual fluxes generated by WRTDS for the nine RIM stations are generally consistent with those generated using ESTIMATOR (fig. 9). Average differences for annual fluxes of total phosphorus range from –17.68 to 12.38 percent (fig. 9). For seven of the nine RIM stations, differences between annual total phosphorus fluxes generated by ESTIMATOR and WRTDS are minimal, ranging from –2.10 to 8.17 percent (all less than 10 percent. The average difference for total phosphorus at the Susquehanna (12.38 percent) (fig. 9A) and Rappahannock (–17.68 percent) (fig. 9D) RIM stations, however, exceeds the 10 percent threshold, which indicates that the annual fluxes generated by WRTDS and ESTIMATOR are different. It has been

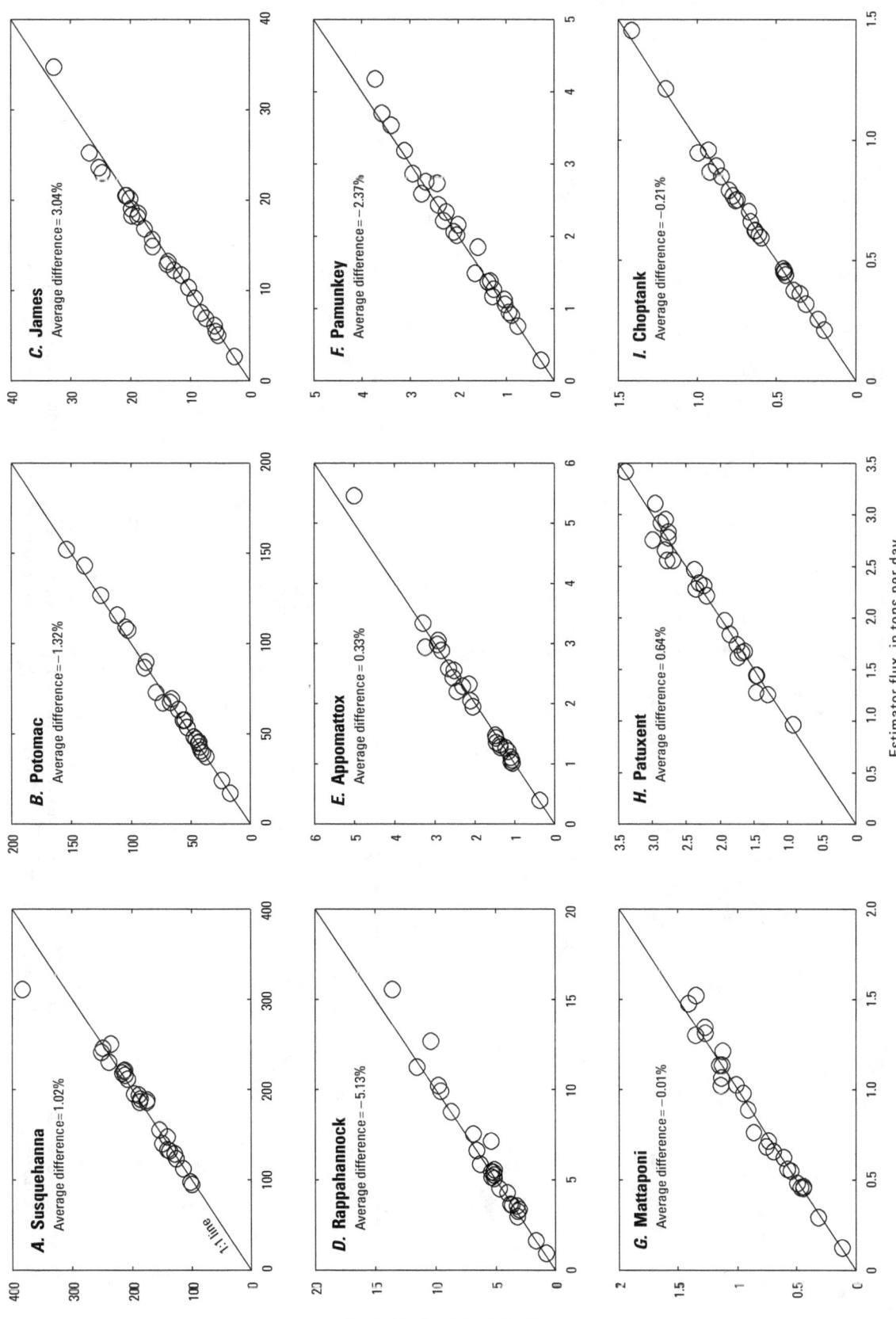

Figure 7. WRTDS-derived versus ESTIMATOR-derived annual total nitrogen fluxes and associated average percent difference at the (A) Susquehanna, (B) Potomac, (C) James, (D) Rappahannock, (E) Appomattox, (F) Pamunkey, (G) Mattaponi, (H) Patuxent, and (I) Choptank River Input Monitoring stations.

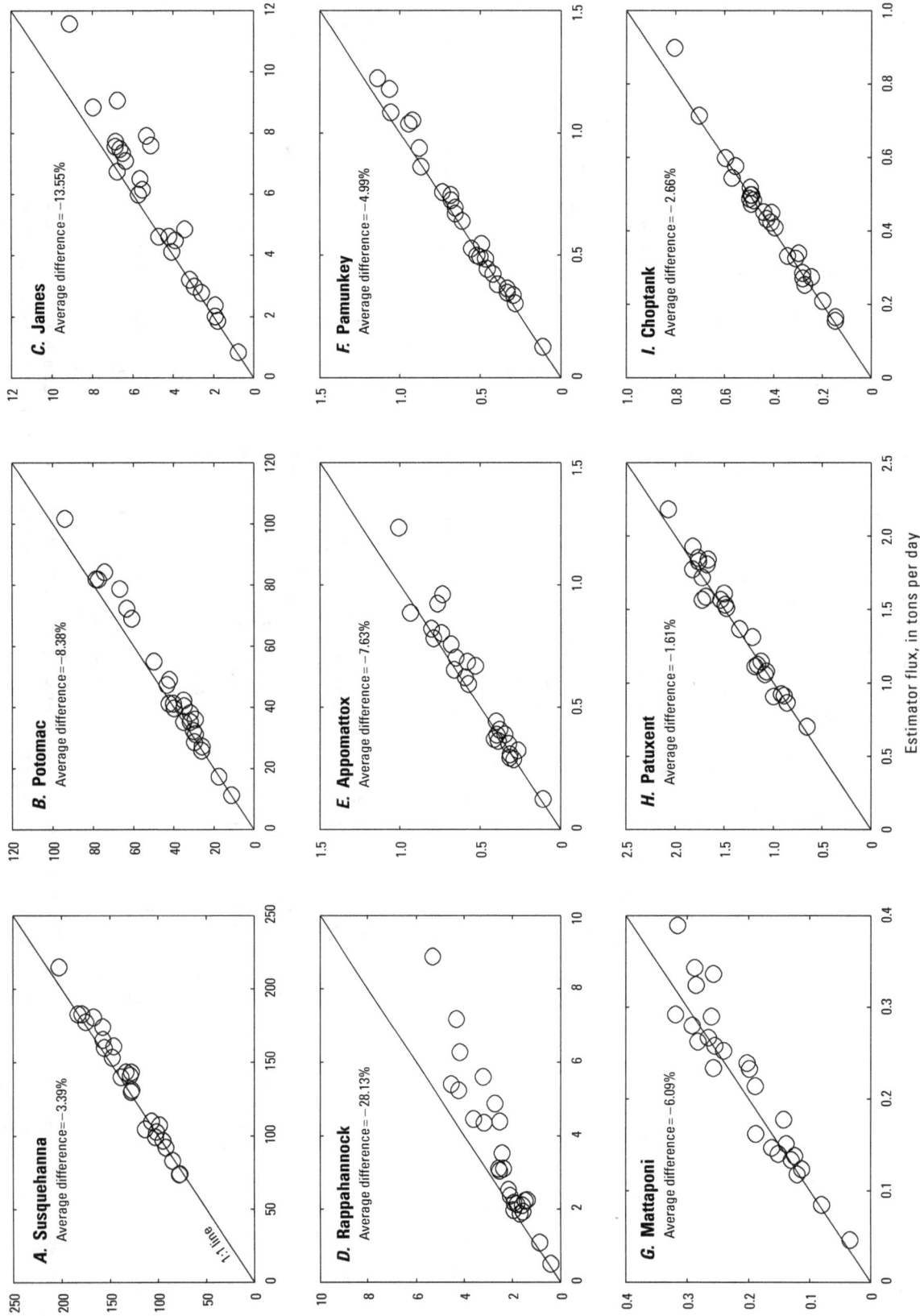

Figure 8.   WRTDS-derived versus ESTIMATOR-derived annual nitrate fluxes and associated average percent difference at the (A) Susquehanna, (B) Potomac, (C) James, (D) Rappahannock, (E) Appomattox, (F) Pamunkey, (G) Mattaponi, (H) Patuxent, and (I) Choptank River Input Monitoring stations.

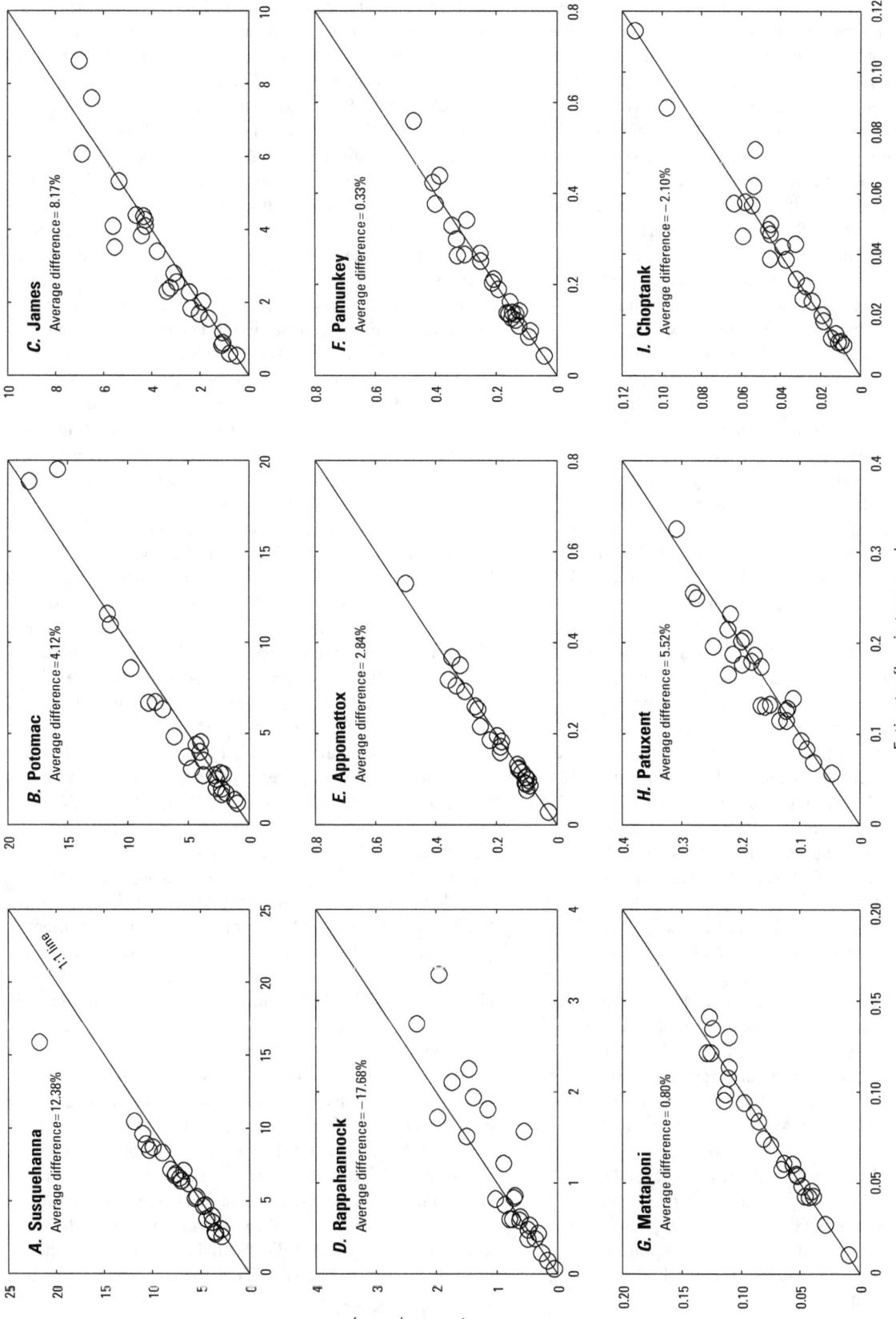

Figure 9. WRTDS-derived versus ESTIMATOR-derived annual total phosphorus fluxes and associated average difference at the (A) Susquehanna, (B) Potomac, (C) James, (D) Rappahannock, (E) Appomattox, (F) Pamunkey, (G) Mattaponi, (H) Patuxent, and (I) Choptank River Input Monitoring stations.

determined that total phosphorus at the Susquehanna station is a Category III combination, which means that WRTDS has a tendency to overestimate concentration and flux at intermediate discharges, and ESTIMATOR tends to underestimate concentration and flux during the highest discharge conditions. The flux-bias ratios (table 2) reveal the cumulative effect of the overestimation by WRTDS and underestimation by ESTIMATOR with ratios of 1.06 and 0.96, respectively. The flux-bias ratios for both models are within 0.10 of 1.0, which indicates both models are minimally biased; however, the total difference between WRTDS and ESTIMATOR flux-bias ratios is 10 percent, which is comparable to the annual flux average difference of 12.38 percent (fig. 9A). The greatest discrepancy in estimated annual fluxes occurred during 2004 when ESTIMATOR predicted an annual flux of 15.9 tons/d and WRTDS predicted an annual flux of 21.7 tons/d. An explanation for this discrepancy can be found in the observed concentration-discharge relations for total phosphorus at the Susquehanna RIM station and how well both WRTDS and ESTIMATOR reproduce this relation (fig. 10). First, ESTIMATOR (fig. 10B) and WRTDS (fig. 10D) appear to reproduce total phosphorus concentration equally well; however, a closer inspection shows that between discharges of approximately 11 and 12.5 log units, WRTDS (black dots) overestimates a greater portion of the observations (red dots) (fig. 10D) compared to ESTIMATOR (fig. 10B). The residuals (observed minus predicted concentration) for ESTIMATOR (fig. 10C) and WRTDS (fig. 10E) predicted concentrations show that WRTDS has a greater density of negative residuals (negative residual means estimated concentration is greater than observed concentration) for discharges between about 11 and 12.5 log units than does ESTIMATOR. The average residual for concentrations in this discharge range is -0.16 (fig. 10C) and -0.22 log units as determined by ESTIMATOR (fig. 10C) and WRTDS (fig. 10D), respectively. Second, ESTIMATOR has a greater tendency to underpredict total phosphorus concentrations/fluxes for extreme high-discharge conditions (greater than 440,000 ft³/s, 13 log units); whereas, WRTDS concentration predictions more closely approximate observed concentrations during these high-discharge conditions (fig. 10B–10E). Annual discharge for the Susquehanna River at the RIM station for 2004 was the highest during the 1985 to 2010 period when the peak daily discharge reached 622,000 ft³/s (13.3 log units). This example shows that the cumulative effect of the over and underprediction of concentration and flux by WRTDS and ESTIMATOR, respectively, is more pronounced in years with greater annual discharge (fig. 9A).

At the Rappahannock station, the average total phosphorus difference of -17.68 percent (fig. 9D) occurs because ESTIMATOR considerably overpredicts total phosphorus concentration associated with discharge conditions greater than 22,000 ft³/s (10 log units) (fig. 11B and 11C); whereas, WRTDS produces estimates of total phosphorus concentration

that more closely reproduce observed concentrations during these high-discharge conditions (fig. 11D and 11E). Observed total phosphorus concentration at the highest discharge (approximately 55,000 ft³/s or 11 log units) is 0.46 mg/L. ESTIMATOR predicted a concentration of 2.72 mg/L and WRTDS predicted a concentration of 0.77 mg/L at this same discharge. This overprediction of concentration by ESTIMATOR during the highest discharges is directly related to Category II discrepancies, as previously discussed, where the observed concentration-discharge relation exhibits a more sinuous pattern with multiple points of inflection. The functional form of the ESTIMATOR model can only represent concentration-discharge relations that are linear or quadratic in form, and the differences between the observed concentration-discharge pattern and a linear or quadratic form cannot be accounted for by the remaining ESTIMATOR variables, time and season.

Orthophosphorus annual fluxes generated by WRTDS for the nine RIM stations are in close agreement with those generated using ESTIMATOR (fig. 12). For eight of the nine RIM stations, the average difference for annual fluxes ranged from -6.85 to 0.17, all of which are considered minimal differences (less than 10 percent); however, the average difference for orthophosphorus at the Patuxent RIM station (10.30 percent) (fig. 12H) just exceeds the 10 percent threshold. The flux-bias ratio results show a very similar pattern in that the flux-bias ratio for WRTDS is 1.09 and the flux-bias ratio for ESTIMATOR is 1.04 (table 2). Both models produce estimates of flux that are positively biased with WRTDS more biased than ESTIMATOR; however, both models are considered minimally biased because both are within 0.10 of 1.0 (no bias). Inspection of each model's performance in reproducing the observed concentration-discharge relation reveals that both models produce similar patterns in orthophosphorus concentration across the full range of flow (fig. 13). Figure13E shows a slight increase in negative residuals (WRTDS predicted concentrations greater than observed concentrations) for low to intermediate discharges that range from approximately 5 to 7 log units (approximately 150 to 1,100 ft³/s); conversely, the ESTIMATOR-derived concentration residuals are more symmetrically distributed around zero for these intermediate-discharge conditions (fig. 13C). This positive bias associated with WRTDS-derived concentrations, however, is diminished (more symmetrical distribution around zero) for discharges greater than approximately 8 log units (approximately 3,000 ft³/s) (fig.13E) compared to ESTIMATOR-derived concentrations residuals (fig. 13C), which are shifted negative within this region of discharge. Therefore, the minor positive shift from the 1:1 line (fig. 12H) may be attributed to the tendency for WRTDS to slightly overpredict orthophosphorus concentrations during intermediate-discharge conditions. This shift is diminished for the highest annual flux, which occurred in 1985, when WRTDS predicted that annual flux was 0.095 tons/d and ESTIMATOR predicted annual flux was 0.088 tons/d (fig.12H).

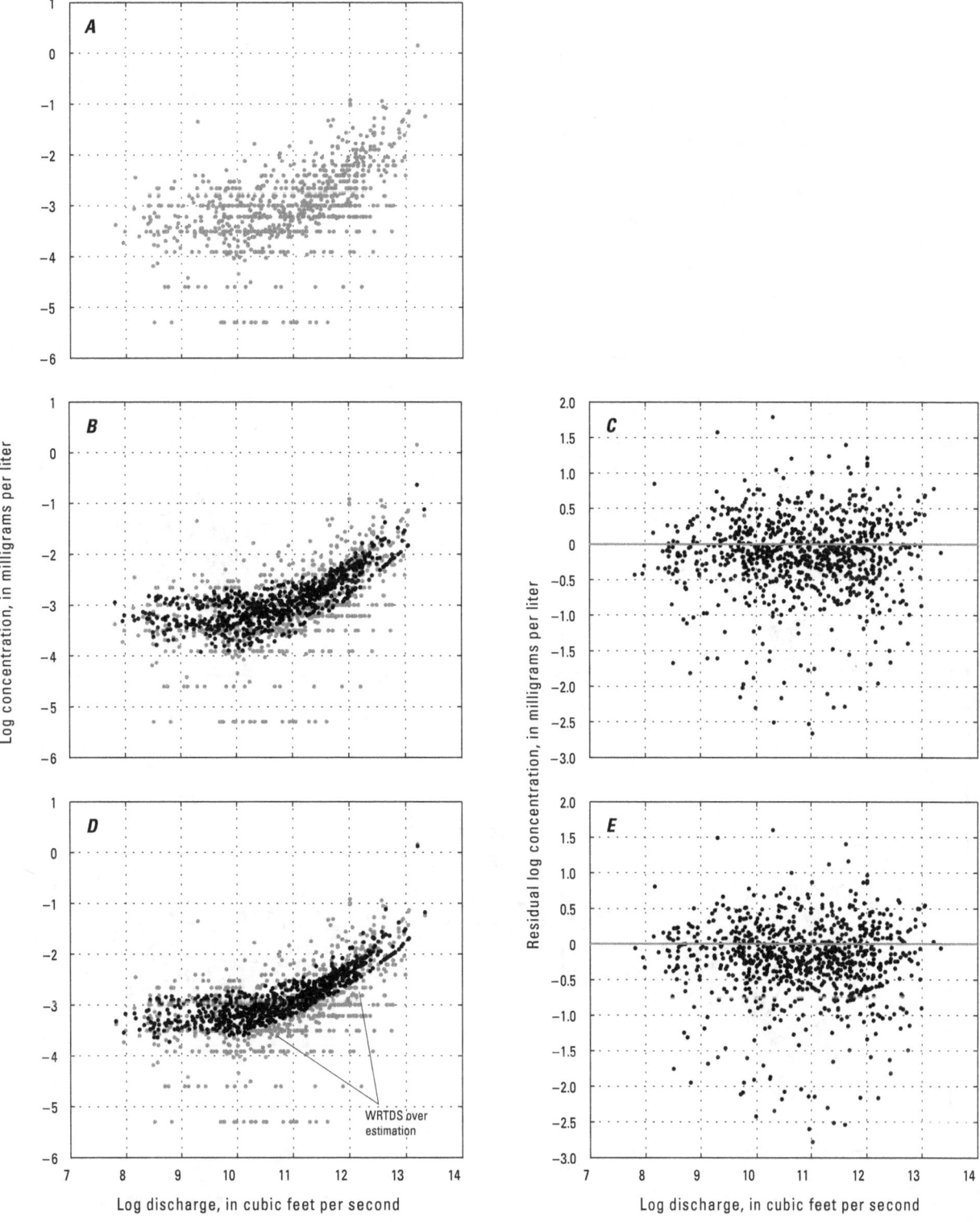

Figure 10. Total phosphorus concentration at Susquehanna River near Conowingo, Maryland (USGS Station ID 01578310), showing the *(A)* observed concentration (red dots) versus discharge relation, *(B)* observed (red dots) and ESTIMATOR-predicted (black dots) concentration versus discharge relation, *(C)* residual (observed minus predicted) plot for ESTIMATOR predictions, *(D)* observed (red dots) and WRTDS-predicted (black dots) concentration versus discharge relation, and *(E)* residual (observed minus predicted) plot for WRTDS predictions.

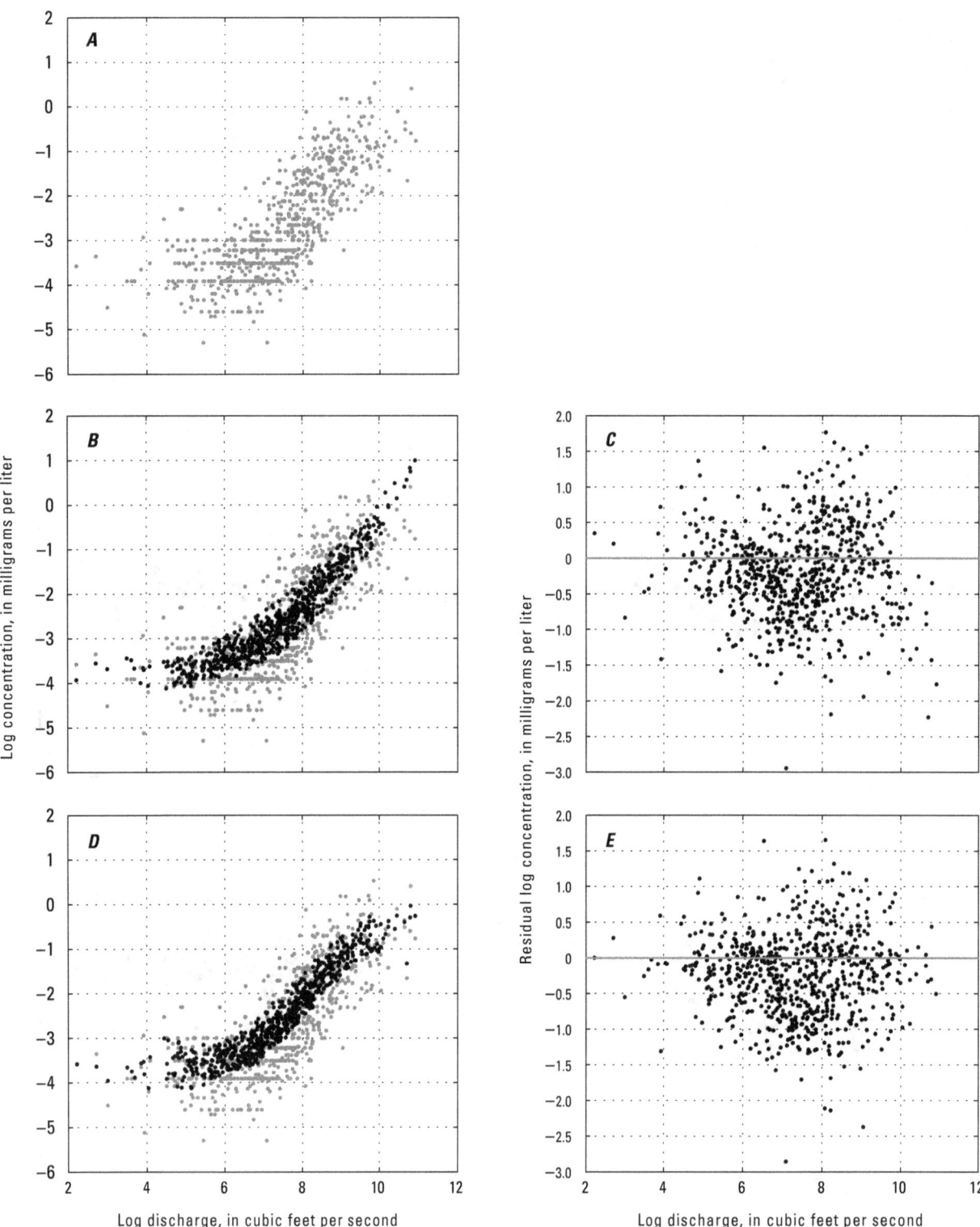

Figure 11.   Total phosphorus concentration at Rappahannock River near Fredericksburg, Virginia (USGS Station ID 01668000), showing the *(A)* observed concentration (red dots) versus discharge relation, *(B)* observed (red dots) and ESTIMATOR-predicted (black dots) concentration versus discharge relation, *(C)* residual (observed minus predicted) plot for ESTIMATOR predictions, *(D)* observed (red dots) and WRTDS-predicted (black dots) concentration versus discharge relation, and *(E)* residual (observed minus predicted) plot for WRTDS predictions.

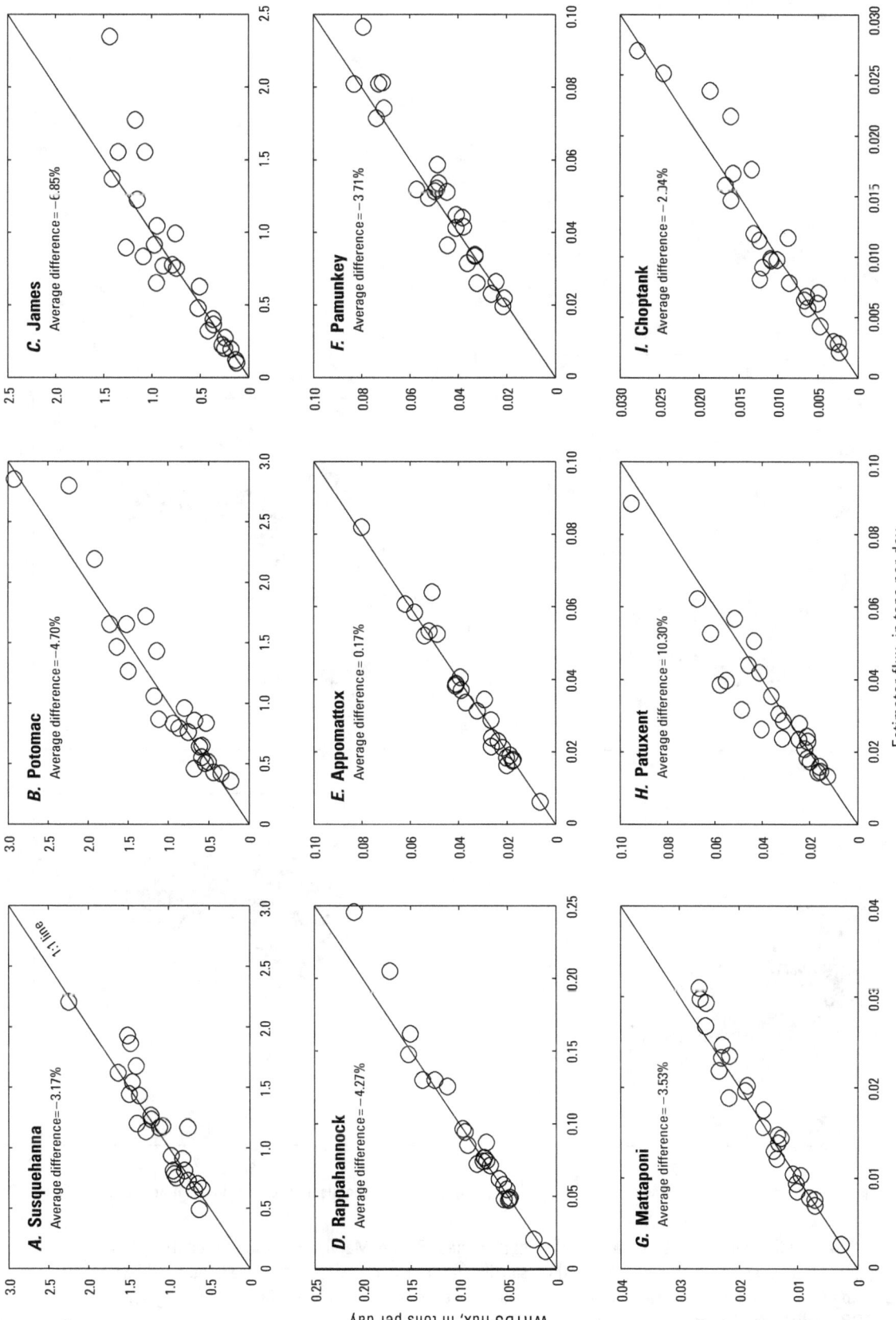

Figure 12. WRTDS-derived versus ESTIMATOR-derived annual orthophosphorus fluxes and associated average difference at the (A) Suscuehanna, (B) Potomac, (C) James, (D) Rappahannock, (E) Appomattox, (F) Pamunkey, (G) Mattaponi, (H) Patuxent, and (I) Choptank River Input Monitoring stations.

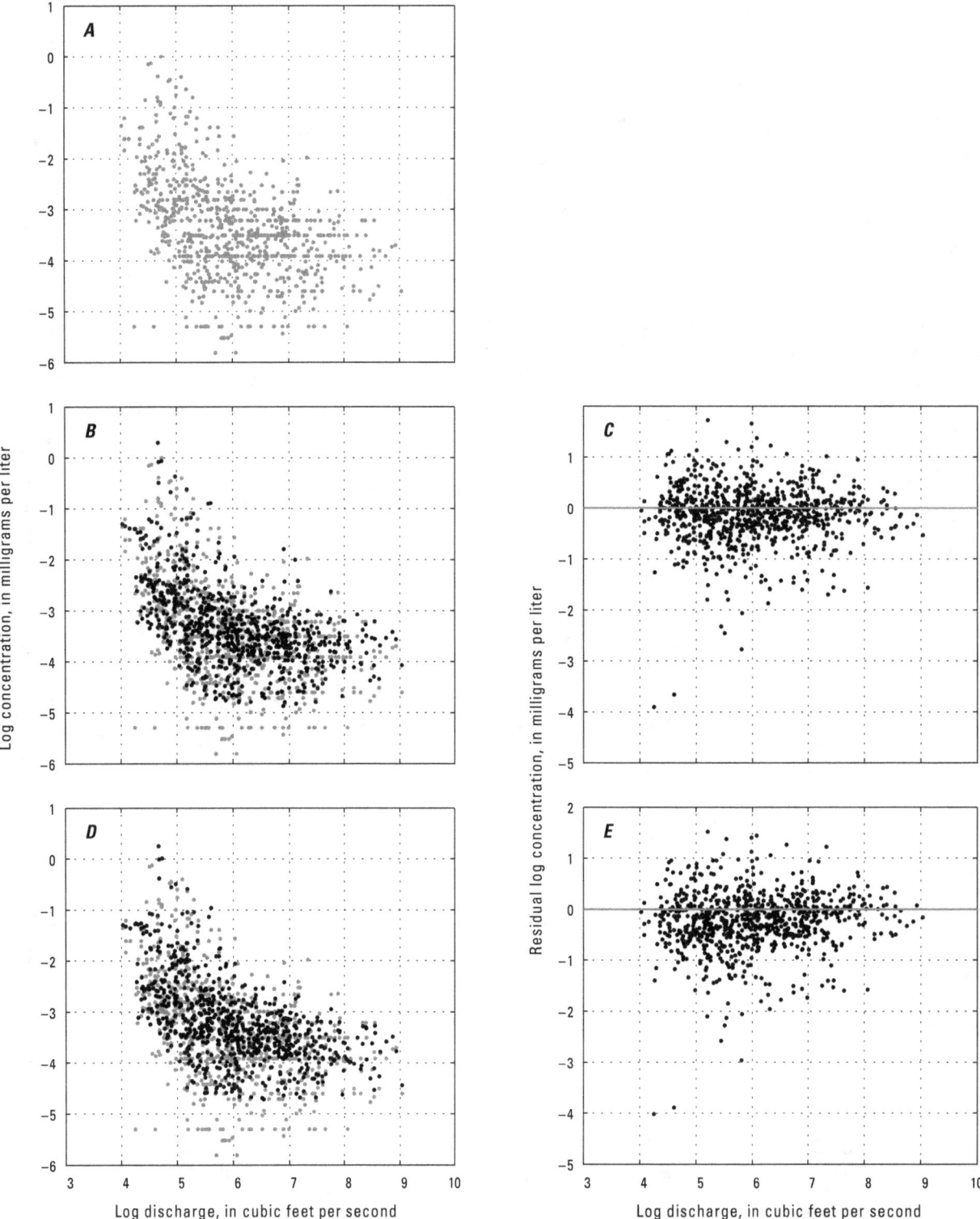

Figure 13.   Orthophosphorus concentration at Patuxent River near Bowie, Maryland (USGS Station ID 01594440), showing the *(A)* observed concentration (red dots) versus discharge relation, *(B)* observed (red dots) and ESTIMATOR-predicted (black dots) concentration versus discharge relation, *(C)* residual (observed minus predicted) plot for ESTIMATOR predictions, *(D)* observed (red dots) and WRTDS-predicted (black dots) concentration versus discharge relation, and *(E)* residual (observed minus predicted) plot for WRTDS predictions.

Suspended Sediment

Suspended sediment annual fluxes generated by WRTDS for the nine RIM stations are, in general, considerably different from those generated using ESTIMATOR (fig. 14). Average differences for annual fluxes of suspended sediment range from –39.50 to 38.33 percent (fig. 14). The average difference for suspended sediment at the nine RIM stations can be split into two groups: (1) stations (eight of nine stations) with negative average percent differences (that is, ESTIMATOR tends to produce higher annual estimates than WRTDS: fig. 14*B*–14*I*) and (2) the Susquehanna RIM station, which has a positive average percent difference (that is, WRTDS tends to produce higher annual estimates than ESTIMATOR: fig. 14*A*). For the first group, average percent difference between WRTDS and ESTIMATOR annual fluxes ranges from –39.50 percent to –2.35 percent. Of the eight stations in this first group, only three have average percent differences greater than 10 percent; these stations are the Potomac (–19.34 percent) (fig. 14*B*), James (–19.20 percent) (fig. 14*C*), and Rappahannock (–39.50 percent) (fig. 14*D*) RIM stations. These three stations are Category II combinations and as a result each has the same root cause for these large negative average percent differences—considerable overprediction of flux by ESTIMATOR during high-discharge conditions. The concentration-discharge relation for suspended sediment at the Rappahannock River RIM station will once again be used to highlight this discrepancy (fig. 4). Figure 4*B* shows that ESTIMATOR (black circles) does a good job reproducing the observed concentration-discharge relation for discharge conditions up to 8,100 ft³/s (9 log units); however, ESTIMATOR considerably overpredicts observed suspended-sediment concentration associated with discharges greater than 8,100 ft³/s. This overprediction, by ESTIMATOR, at the Rappahannock, James, and Potomac RIM stations is directly is related to Category II discrepancies, previously discussed, where the observed concentration-discharge relation exhibits a more sinuous pattern with multiple points of inflection. The functional form of the ESTIMATOR model can only represent concentration-discharge relations that are linear or quadratic in form, and the differences between the observed concentration-discharge pattern and a linear or quadratic form cannot be accounted for by the remaining ESTIMATOR variables, time and season. These three stations exhibit a sigmoidal pattern in the observed concentration-discharge relation that can only be partially reproduced by ESTIMATOR (fig. 4*B*); conversely, WRTDS, which has a more flexible design, is able to reproduce the full sigmoidal pattern of the observed concentration-discharge relation (fig. 4*D*). At the Susquehanna station, two issues are contributing to the elevated average percent difference in suspended sediment annual fluxes (38.33 percent). First, WRTDS has a tendency to overpredict suspended sediment flux for intermediate- to high-discharge conditions ranging from 22,000 to 163,000 ft³/s (10 to 12 log units) (fig. 6*D* and 6*E*); whereas, ESTIMATOR produces more accurate estimates in this range of flow conditions (fig. 6*B* and 6*C*). This discharge condition typically dominates the annual hydrographs given that over 95 percent of the daily discharges, occurring between 1985 and 2010, are less than 163,000 ft³/s (approximately 12 log units). Therefore, for the majority of years, the annual flux is generated from discharge conditions that extend up to 163,000 ft³/s, and the positive shift from the 1:1 line (fig. 14*A*) for the range extending from 5,000 to 10,000 tons/d can primarily be attributed to the overestimation by WRTDS during these intermediate-high discharge conditions. Second, ESTIMATOR has a greater tendency to considerably underpredict suspended-sediment concentrations/fluxes for extreme high-discharge conditions (that is, greater than 440,000 ft³/s, 13 log units) (fig. 6*B* and 6*C*); whereas, WRTDS flux predictions are more accurate during these discharge conditions (fig. 6*D*–6*E*). The greatest Susquehanna total phosphorus annual flux estimated by both WRTDS (34,721 tons/d) and ESTIMATOR (13,007 tons/d) occurred in 2004, which was the wettest year for the period 1985 to 2010. The deviation from the 1:1 line (fig. 7*A*) is more a result of ESTIMATOR tending to underpredict total phosphorus concentrations/fluxes associated with extreme high-flow conditions (greater than 440,000, ft³/s, 13 log units) than the positive bias of WRTDS estimates during intermediate-discharge conditions.

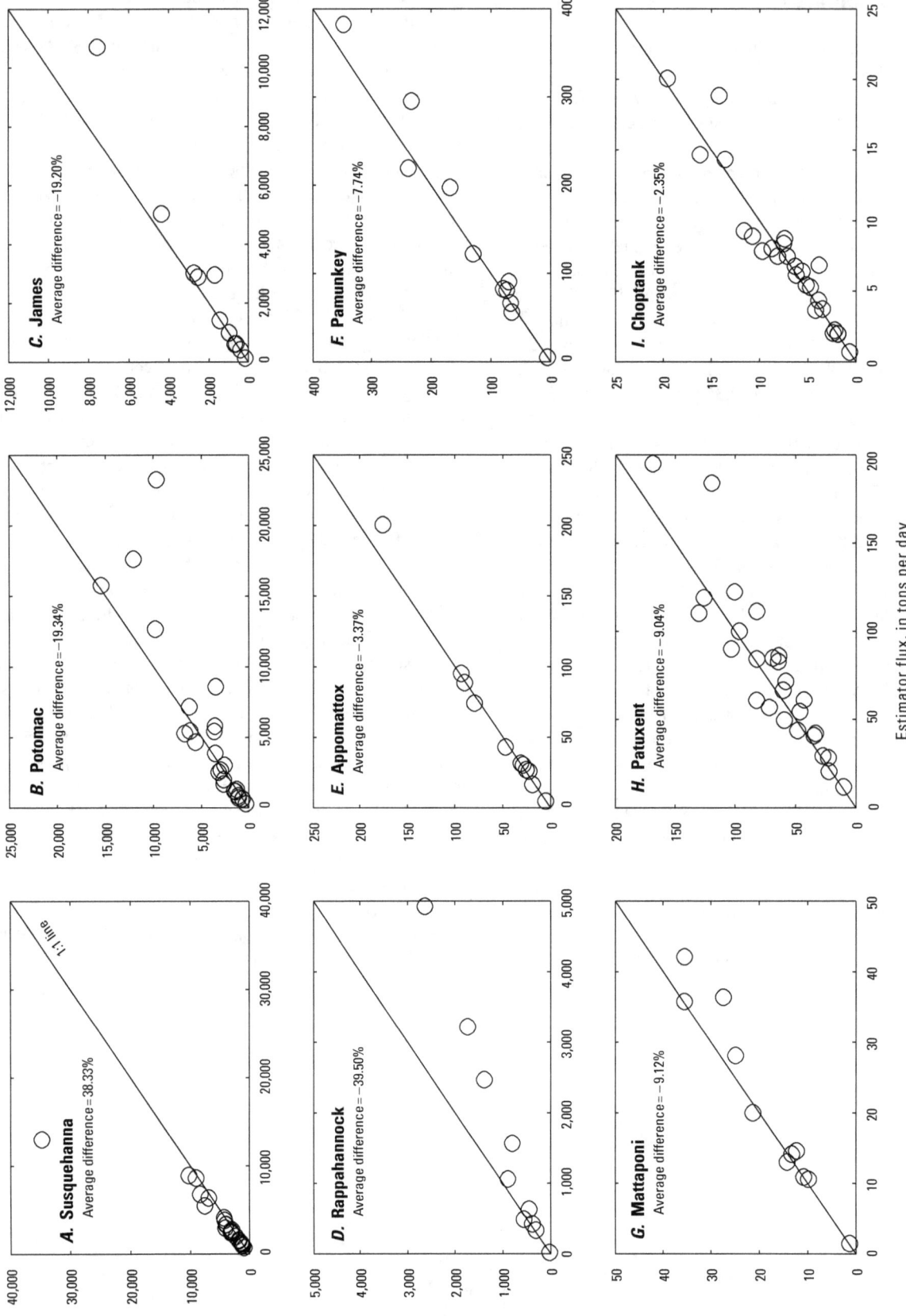

Figure 14. WRTDS-derived versus ESTIMATOR-derived annual suspended-sediment fluxes and associated average percent difference at the (*A*) Susquehanna, (*B*) Potomac, (*C*) James, (*D*) Rappahannock, (*E*) Appomattox, (*F*) Pamunkey, (*G*) Mattaponi, (*H*) Patuxent, and (*I*) Choptank River Input Monitoring stations.

Trends in Nutrient and Sediment Flux to the Chesapeake Bay

For the presentation of changes in nutrient and sediment flux, two perspectives are provided. First, the WRTDS flow-normalized changes in flux (table 3) will be compared to the historical trends in flow-adjusted concentration (table 3) obtained from ESTIMATOR. Second, the way in which nutrient and sediment fluxes are changing in each of the nine major Chesapeake Bay tributaries will be discussed. Changes in flow-normalized nutrient and sediment flux are reported for two time periods: 1985 to 2010 and 2001 to 2010. These time periods are used because 1985 is viewed by the CBP as the base year that all subsequent years are compared in order to assess changes in nutrient and sediment concentrations and now fluxes, and 2001 serves as the beginning year for the most recent 10-year period. The CBP requests that results be provided for both long-term (since 1985) and short-term (last 10 years) trends to

assess the influence of management activities on nutrient and sediment transport throughout the Chesapeake Bay watershed. Historically, analysis of 10 years of data was the shortest period for which ESTIMATOR could determine flow-adjusted concentration trends and minimizing associated uncertainty. WRTDS has the ability to determine trends for any specified time period for records of at least 20 years. The slope defining flux changes at each of the nine RIM stations can be reported as (1) a percent change per year (eq. 6) and (2) a mass change per year (eq. 7). Because the range in drainage areas for the nine RIM stations extends from 113 square miles (mi^2) for the Choptank River to 27,100 mi^2 for the Susquehanna River, yield (flux per unit of drainage area and expressed as pounds per day per square mile) is used instead of flux so that yield and changes in yield comparison can be made among all nine RIM stations. Plots of annual and flow-normalized flux, tables of changes in annual flux, and time series of annual and flow-normalized flux for each of the nine RIM stations are provided in appendixes 2, 3, and 4, respectively.

Table 3. Total nitrogen trends for WRTDS flow-normalized yields and ESTIMATOR flow-adjusted concentration at the nine River Input Monitoring (RIM) stations for the time periods 1985 to 2010 and 2001 to 2010.

[%, percent; %/yr, percent per year; WRTDS flow-normalized yield results are presented as both the total percent change in yield and average slope (percent change in yield per year); ESTIMATOR changes in flow-adjusted concentration are reported as a 95-percent confidence interval (CI) for the average change in concentration per year; in the two ESTIMATOR columns of the table, shaded cells are those where the ESTIMATOR flow-adjusted concentration trend is significant, those with no shading are not significant. Pink shaded cells indicate that the ESTIMATOR flow-adjusted concentration trend and the WRTDS flow-normalized yield trend have the same sign; black text indicates that the WRTDS trend in flow-normalized yield falls within the 95-percent confidence interval for ESTIMATOR flow-adjusted concentration; red text indicates that the WRTDS trend in flow-normalized yield falls outside the 95-percent confidence interval for ESTIMATOR flow-adjusted concentration; all values are rounded to the nearest tenth]

RIM station	WRTDS flow-normalized yield				ESTIMATOR flow-adjusted concentration	
	1985 to 2010		2001 to 2010		1985 to 2010	2001 to 2010
	Total change (%)	Slope (%/yr)	Total change (%)	Slope (%/yr)	Slope (Range of 95% CI, %/yr)	Slope (Range of 95% CI, %/yr)
Susquehanna	−20.8	−0.8	−5.8	−0.6	−1.2 to −0.8	−1.0 to 1.4
Potomac	−14.3	−0.6	−3.6	−0.4	−1.1 to −0.8	−2.0 to −0.2
James	−8.2	−0.3	6.5	0.7	−1.1 to −0.3	−1.8 to 1.2
Rappahannock	−8.5	−0.3	−4.3	−0.5	−0.8 to 0.1	−3.0 to 0.9
Appomattox	−3.5	−0.1	3.3	0.4	−0.1 to 0.6	−1.1 to 1.4
Pamunkey	2.8	0.1	4.9	0.5	0.4 to 1.2	−1.4 to 1.0
Mattaponi	−4.2	−0.2	4.8	0.5	−0.3 to 0.3	−1.2 to 0.6
Patuxent	−49.3	−2.0	−10.6	−1.2	−2.4 to −2.2	−3.0 to −1.4
Choptank	7.4	0.3	7.4	0.8	0.0 to 0.5	−0.5 to 1.6

Comparison of ESTIMATOR and WRTDS Trend Results

To effectively present the new WRTDS results, information is provided on how similar or different the WRTDS trends in flow-normalized flux are compared to the historical trends in flow-adjusted concentration, derived from ESTIMATOR, and explanation for major differences in trend (as determined by changes in trend direction) is also provided. The problem with comparing changes in flow-normalized flux (WRTDS) to changes in flow-adjusted concentration (ESTIMATOR) is related to their respective limitations. ESTIMATOR's assumption of time-invariance in the relation between constituent concentration and discharge, time, and season inherent in the single model window approach is not appropriate for all constituents at the nine RIM stations (as documented earlier in the comparison of methods for the determination of flux). Also, the flow-adjusted concentration trend from ESTIMATOR may not be representative of the actual trends in flux. Two limitations with the WRTDS approach are the lack of measurements of uncertainty associated with trend (for example, 95-percent confidence interval) and the inability to assign significance to the trend (that is, determine if the trend is significantly different from zero).

ESTIMATOR flow-adjusted concentration and WRTDS flow-normalized flux trends were compared to answer two questions: (1) Does the WRTDS flow-normalized flux trend fall within the 95-percent confidence interval associated with the ESTIMATOR flow-adjusted concentration trend? and (2) Is the direction of change (that is, improving or degrading conditions) consistent between flow-normalized flux and flow-adjusted concentration for stations and constituents that have a significant (p-value < 0.05) flow-adjusted concentration trend? For this comparison, there are 85 constituent/station combinations that have trend information for the two time periods (the reason for 85 instead of 90 combinations is that the five RIM stations in Virginia do not have trend information for suspended sediment for the period 1985 to 2010). For 55 of 85 possible combinations (64 percent), the WRTDS flow-normalized flux (in percent change per year) resides within the 95-percent confidence interval for the ESTIMATOR flow-adjusted concentration trend (in percent change per year) (bold black text, tables 3–7). For 48 of the 85 combinations, the ESTIMATOR flow-adjusted concentration trend is significantly different from zero (shaded cells, tables 3–7) (historically, the magnitude of flow-adjusted concentration trends was reported only for combinations with significant trends; whereas, the magnitude was not reported for combinations with non-significant trends (Langland and others, 2006)). There is agreement between flow-normalized flux and flow-adjusted concentration, with respect to the direction of change (that is, improving or degrading conditions), at 38 of the 48 combinations (79 percent) with significant flow-adjusted concentration trends (pink shaded cells, tables 3–7). Ten of the 48 combinations (21 percent) have trend results where WRTDS flow-normalized

flux trends are in an opposite direction when compared to ESTIMATOR flow-adjusted concentration trend (blue shaded cells, tables 3–7). So, the question becomes why are there cases of differences in the trend directions between WRTDS flow-normalized flux and ESTIMATOR flow-adjusted concentration? To answer this question, an investigation was conducted to determine how observed concentrations are changing (presumably as a result of management activities) within different portions of the complete range of hydrologic conditions. Three examples are provided to illustrate when (1) the direction of the flow-normalized flux and flow-adjusted concentration trends agree; (2) flow-normalized flux trends are positive (degrading conditions) and flow-adjusted concentration trends are negative (improving conditions); and (3) flow-normalized flux trends are negative (improving conditions) and flow-adjusted concentration trends are positive (degrading conditions). For each of these examples, observed concentrations were separated into three discharge intervals: discharges less than or equal to the 60th percentile (base-flow conditions); discharges that are greater than the 60th percentile but less than the 90th percentile (intermediate discharges); and discharges greater than the 90th percentile (high discharges). The first example is for combinations where the trend in flow-adjusted concentration is in the same direction as the trend in flow-normalized flux, which accounts for 38 of the 48 combinations that have significant ESTIMATOR flow-adjusted concentration trends. Figure 15 shows total nitrogen collected at the Susquehanna River RIM station and is an example of this category. The figure shows that total nitrogen concentrations collected during high (discharge greater 11.32 log units or 82,700 ft^3/s) (fig. 15A), intermediate (discharge between 9.96 and 11.32 log units or 21,200 to 82,700 ft^3/s) (fig. 15B), and low (less than 9.96 log units or 21,200 ft^3/s) (fig. 15C) discharges all exhibit patterns of decreasing concentrations from 1985 to 2010 (the red line is a loess smooth fit line to serve as a visual aid for detecting direction of change and not a function of either WRTDS or ESTIMATOR). The WRTDS flow-normalized flux for the period 1985 to 2010 is -0.8 percent per year, and the 95-percent confidence interval for the flow-adjusted concentration trend is -1.2 to -0.8 percent (table 3). Because the slopes in total nitrogen concentration are similar for all three discharge categories, the flow-normalized flux and flow-adjusted concentration trends are statistically indistinguishable. The second example shows total phosphorus collected at the James River RIM station and is representative of 7 of 48 combinations where the trend in WRTDS flow-normalized flux is positive (degrading condition) and the trend in ESTIMATOR flow-adjusted concentration is negative (improving condition) (fig. 16). Figure 16C shows that total phosphorus concentrations associated with low-flow/base-flow conditions have decreased sharply during the period from 1985 to 2010. This pattern also is evident but less pronounced for concentrations associated with intermediate discharges (fig. 16B); however, total phosphorus concentrations associated with high discharges (fig. 16A) show a gradual increase starting about 2000. The effect of these patterns is evident in the flux

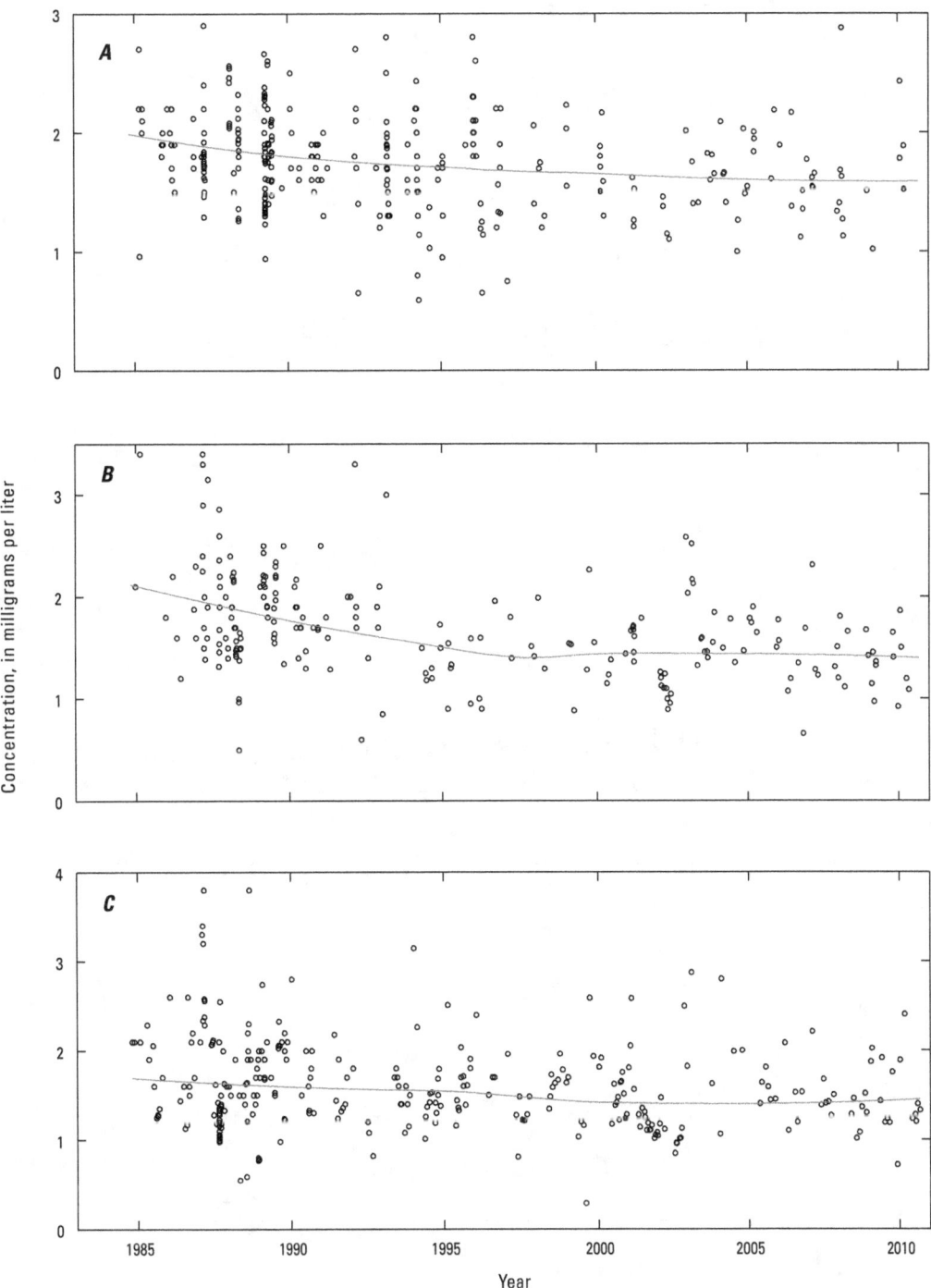

Figure 15. Total nitrogen at the Susquehanna River at Conowingo, Maryland (USGS Station ID 01578310), showing observed concentration collected during *(A)* high (greater than the 90th percentile of discharge), *(B)* intermediate (greater than the 60th and less than the 90th percentile of discharge), and *(C)* low (less than or equal to the 60th percentile of discharge) discharges. Red line represents the Loess smooth fit line used as a visual aid for changing concentrations and not associated with either WRTDS or ESTIMATOR.

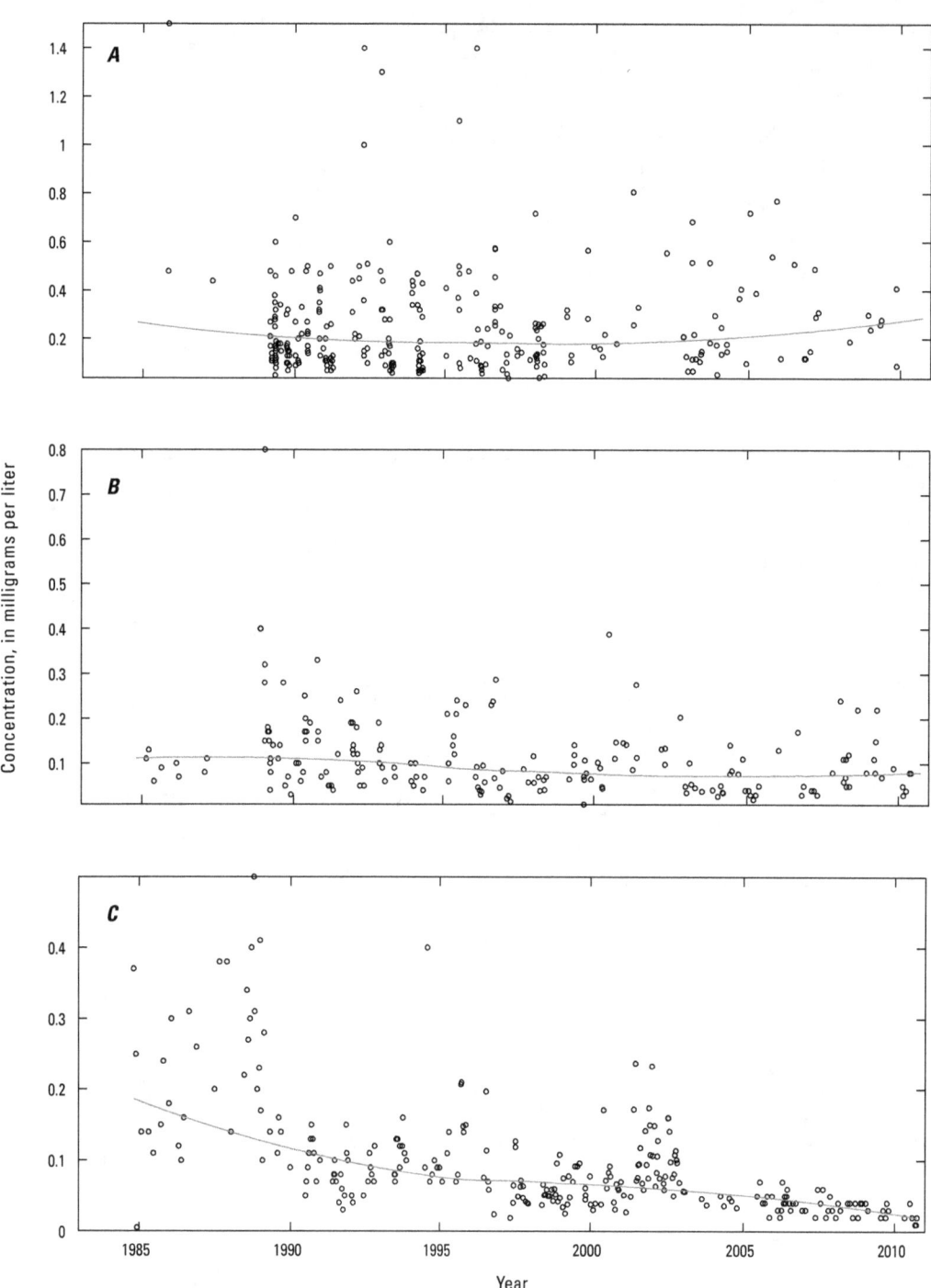

Figure 16. Total phosphorus at the James River at Cartersville, Maryland (USGS Station ID 02035000), showing observed concentration collected during *(A)* high (greater than the 90th percentile of discharge), *(B)* intermediate (greater than the 60th and less than the 90th percentile of discharge), and *(C)* low (less than or equal to the 60th percentile of discharge) discharges. Red line represents the Loess smooth fit line used as a visual aid for changing concentrations and not associated with either WRTDS or ESTIMATOR.

and concentration trends where WRTDS flow-normalized flux trend shows an increase of 0.4 percent per year whereas the ESTIMATOR flow-adjusted concentration trend has a 95-percent confidence interval showing a decrease of −2.7 to −2.2 percent per year (table 5). These trends and patterns in the observed concentration suggest that management efforts (primarily upgrades to municipal wastewater-treatment plants and the phosphorus detergent ban) to reduce total phosphorus in the James River watershed are most evident in concentrations associated with low discharges. Conversely, the increasing trend in flow-normalized flux and observed concentration associated with high discharges indicates that management activities have not had a net effect in reducing total phosphorus loading from nonpoint sources. The last example shows nitrate collected at the Pamunkey RIM station and is representative of 3 of 48 combinations where the trend in WRTDS flow-normalized flux is negative (improving condition) and the trend in ESTIMATOR flow-adjusted concentration is positive (degrading condition) (fig. 17). Figure 17C shows that nitrate concentrations associated with base-flow conditions increased in the overall period 1985 to 2010 with the greatest rate of increase occurring between

1985 and 1999. For intermediate flows, nitrate concentrations increased between 1985 and 2002 and decreased from 2003 to 2010 (fig. 17B). Conversely, nitrate concentrations associated with high discharges show a pattern of gradually decreasing concentrations between 2002 and 2010 (fig. 17A). The effect of these patterns is evident in the flux and concentration trends where WRTDS flow-normalized flux trend shows a gradual decrease of −0.3 percent per year whereas the ESTIMATOR flow-adjusted concentration trend has a 95-percent confidence interval showing a more pronounced increase of 0.8 to 2.1 percent per year (table 4). These three examples show that (1) ESTIMATOR flow-adjusted concentration trends are heavily influenced by the patterns in constituent concentration during low and intermediate discharges, whereas WRTDS flow normalized fluxes are heavily influenced by the patterns in constituent concentrations associated with high discharges; and (2) the WRTDS flow-normalized flux trends and the ESTIMATOR flow-adjusted concentration trends are correctly characterizing changes in water-quality conditions specific to changes in flux and concentration, respectively and there are times that trends in flux will be in the opposite direction of trends in concentration.

Table 4. Nitrate trends in WRTDS flow-normalized yields and ESTIMATOR flow-adjusted concentration at the nine River Input Monitoring (RIM) stations for the time periods 1985 to 2010 and 2001 to 2010.

[%, percent; %/yr, percent per year; WRTDS flow-normalized yield results are presented as both the total percent change in yield and average slope (percent change in yield per year); ESTIMATOR changes in flow-adjusted concentration are reported as a 95-percent confidence interval (CI) for the average change in concentration per year; in the two ESTIMATOR columns of the table, shaded cells are those where the ESTIMATOR flow-adjusted concentration trend is significant, those with no shading are not significant. Pink shaded cells indicate that the ESTIMATOR flow-adjusted concentration trend and the WRTDS flow-normalized yield trend have the same sign; blue shaded cells indicate that the ESTIMATOR flow-adjusted concentration trend and the WRTDS flow-normalized yield trend have opposite signs. Black text indicates that the WRTDS trend in flow-normalized yield falls within the 95-percent confidence interval for ESTIMATOR flow-adjusted concentration; red text indicates that the WRTDS trend in flow-normalized yield falls outside the 95-percent confidence interval for ESTIMATOR flow-adjusted concentration; using this coding scheme, the most substantial apparent contradictions are those in the blue shaded cells (significant ESTIMATOR trend and opposite signs to the trend directions); all values are rounded to the nearest tenth]

| RIM station | WRTDS flow-normalized yield | | | | ESTIMATOR flow-adjusted concentration | |
| | 1985 to 2010 | | 2001 to 2010 | | 1985 to 2010 | 2001 to 2010 |
	Total change (%)	Slope (%/yr)	Total change (%)	Slope (%/yr)	Slope (Range of 95% CI, %/yr)	Slope (Range of 95% CI, %/yr)
Susquehanna	−14.9	−0.6	−11.5	−1.3	−0.9 to −0.5	−1.2 to 1.6
Potomac	−20.9	−0.8	−18.9	−2.1	−1.5 to −1.0	−3.2 to −0.4
James	−20.8	−0.8	−7.3	−0.8	−1.7 to −0.7	−1.2 to 8.3
Rappahannock	−18.6	−0.7	−14.8	−1.6	−1.5 to 0.1	−2.6 to 10.0
Appomattox	−13.2	−0.5	−8.3	−0.9	−1.2 to 0.3	−1.5 to 7.3
Pamunkey	−6.5	−0.3	−16.5	−1.8	0.8 to 2.1	−2.0 to 1.5
Mattaponi	−6.3	−0.3	−4.9	−0.5	−0.3 to 1.2	−2.4 to 4.1
Patuxent	−49.1	−2.0	−13.3	−1.5	−2.4 to −2.2	−3.1 to −1.2
Choptank	33.0	1.3	9.8	1.1	1.0 to 1.9	−0.7 to 2.3

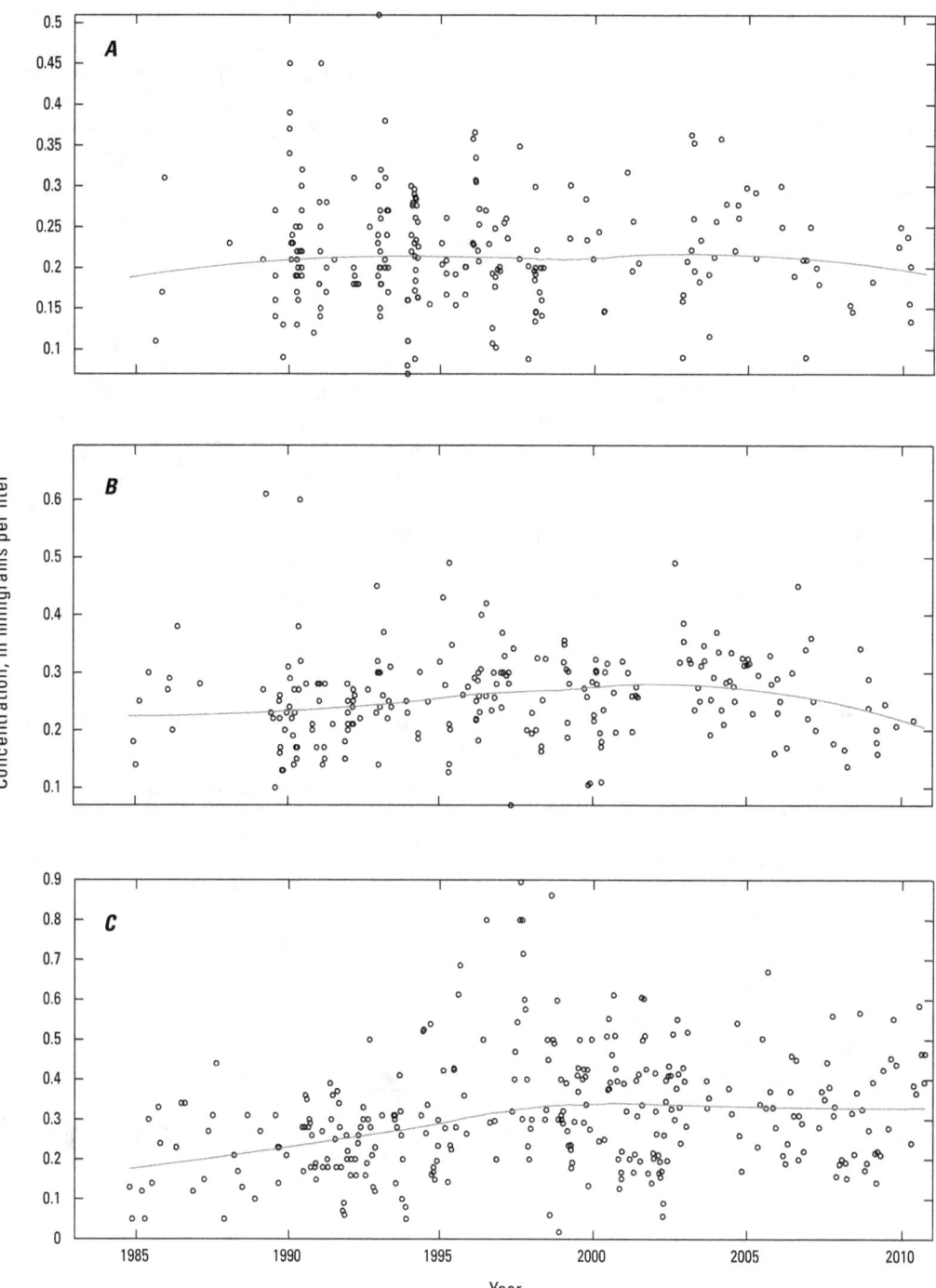

Figure 17.   Nitrate at the Pamunkey River near Hanover, Virginia (USGS Station ID 01673000), showing observed concentration collected during *(A)* high (greater than the 90th percentile of discharge), *(B)* intermediate (greater than the 60th and less than the 90th percentile of discharge), and *(C)* low (less than or equal to the 60th percentile of discharge) discharges. Red line represents the Loess smooth fit line used as a visual aid for changing concentrations and not associated with either WRTDS or ESTIMATOR.

Changes in Flow-Normalized Flux at Individual Stations

The focus for the remainder of this section will be to report changes in WRTDS flow-normalized flux for total nitrogen, nitrate, total phosphorus, orthophosphorus, and suspended sediment at the nine RIM stations for the periods 1985 to 2010 and 2001 to 2010.

Nitrogen

Annual average (black circles) and flow-normalized annual average (red line) total nitrogen yield for the nine RIM stations are presented in figure 18 for the period 1985 to 2010. Total nitrogen yield at the majority of the nine RIM stations commonly ranges between 5 and 20 pounds per day per square mile [(lb/d)/mi^2]. However, the Appomattox, Pamunkey, and Mattaponi RIM stations exhibit yields that are the lowest of the nine RIM stations and are routinely less than 5 (lb/day)/mi^2 (fig. 18*E*, *F*, and *G*). The Pamunkey and Mattaponi Rivers are low-gradient and primarily forested watersheds that reside almost entirely in the Coastal Plain Physiographic Province (fig. 1). These basin characteristics function to limit total nitrogen transport by limiting contributing nitrogen sources and increasing residence time, which promotes greater rates of biological processing and uptake. Not only is the Appomattox River a primarily forested watershed, but more importantly, it is located directly downstream from the Lake Chesdin reservoir, which limits the downstream transport of total nitrogen because of impounded sediment. The extent to which annual yields have changed outside the year-to-year variations in discharge is provided in table 3 for two time periods, 1985 to 2010 and 2001 to 2010. This change in total nitrogen flux is reported, in table 3 (in the WRTDS columns), as a total change in percent and as a percentage change per year. The rate of change in total nitrogen flux (and all subsequent constituents) for each RIM station reported as tons per day per year is provided in Appendixes 3–1 through 3–5. For the period 1985 to 2010, seven of the nine RIM stations had negative slopes, indicating a decrease in the annual delivery of total nitrogen to the Chesapeake Bay. The three stations with the greatest improvement in total nitrogen are the Patuxent (−2.0 percent per year, total reduction of −49.3 percent), Susquehanna (−0.8 percent per year, total reduction of −20.8 percent), and the Potomac (−0.6 percent per year, total reduction of −14.3 percent). Only two stations had increasing yields over the 1985 to 2010 period: the Choptank (0.3 percent per year; total increase of 7.4 percent) and the Pamunkey (0.1 percent per year, total increase of 2.8 percent). For the period 2001 to 2010, only the Susquehanna (−0.6 percent per year), Potomac (−0.4 percent per year), Rappahannock (−0.5 percent per year) and Patuxent (−1.2 percent per year) stations showed a continued reduction (negative slopes); however, the rate of improvement has

decreased at the Susquehanna, Potomac, and Patuxent stations compared to the rates during 1985 to 2010. The Rappahannock is the only station where the rate of improvement has increased. The James, Appomattox, and Mattaponi stations had negative slopes during 1985 to 2010 that have become positive for the period 2001 to 2010. The positive slope in total nitrogen yields at the Choptank (0.3 percent per year) and Pamunkey (0.1 percent per year) stations during 1985 to 2010 steepened during the more recent period of 2001 to 2010 to 0.8 and 0.5 percent per year, respectively (table 3). These results for total nitrogen yield tell two very different stories. Flow-normalized trends in total nitrogen yields indicate that for the long-term period all but two of the RIM stations have total nitrogen yields that are improving. Conversely, flow-normalized trends in total nitrogen yield at eight of the nine RIM stations show that the rate of improvement has either slowed (three of eight RIM stations), changed from improving to degrading (three of eight RIM stations), or the rate of degradation has accelerated (two of the eight RIM stations) during the more recent 2001 to 2010 period.

Annual average and flow-normalized annual average nitrate yields for the nine RIM stations are presented in figure 19 for the period 1981 to 2010. The James, Rappahannock, Appomattox, Pamunkey, and Mattaponi stations have nitrate yields that are noticeably lower than the remaining four RIM stations and routinely are below 5 (lb/d)/mi^2. Conversely, the Susquehanna, Potomac, Patuxent, and Choptank stations have the highest yields that commonly range between 5 and 15 (lb/d)/mi^2. Data in table 4 indicate that nitrate yields are decreasing (negative slopes) at all RIM stations except for the Choptank station for the period 1985 to 2010. These negative slopes range from −2.0 to −0.3 percent per year. The greatest reductions in nitrate yield occurred at the Patuxent (−2.0 per year, total reduction of 49.1 percent), Potomac (−0.8 percent per year, total reduction of 20.9 percent), James (−0.8 percent per year, total reduction of 20.8 percent), and Rappahannock (−0.7 percent per year, total reduction of 18.6 percent). At the Susquehanna, Potomac, Rappahannock, Appomattox, Pamunkey, and Mattaponi stations, the negative slopes in nitrate yield observed during 1985 to 2010 steepened considerably (greater rates of improvement) during the period of 2001 to 2010 (table 4). The Choptank is the only RIM station that had a positive slope for annual-average yield; the positive slope for 1985 to 2010 was 1.3 percent per year (total increase of 33.0 percent), which is greater than the 1.0 percent per year slope observed during the 2001 to 2010. These results show that nitrate yields are improving at eight of the nine RIM stations and that during the 2001 to 2010 time period the rate of improvement has increased at seven of the nine RIM stations. The flow-normalized trend results for nitrate yield are considerably different than the trends for total nitrogen during the period 2001 to 2010 in that nitrate is showing patterns of increased improvement; whereas, trends in total nitrogen yield indicate that the rate of improvement is slowing or becoming increasingly degraded.

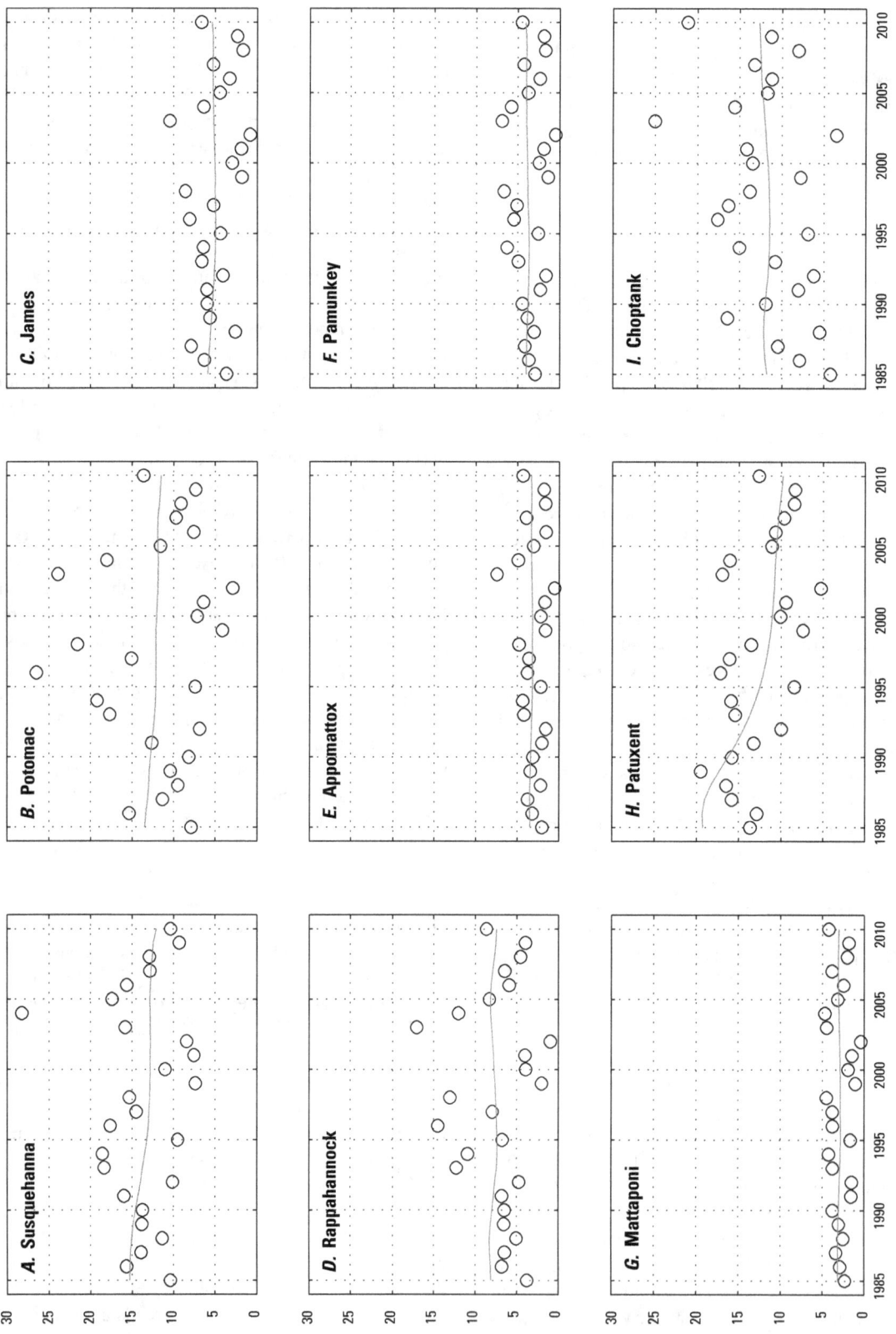

Figure 18. Total nitrogen yields (black circles) and flow-normalized yields (red line) for the nine RIM stations using the WRTDS method.

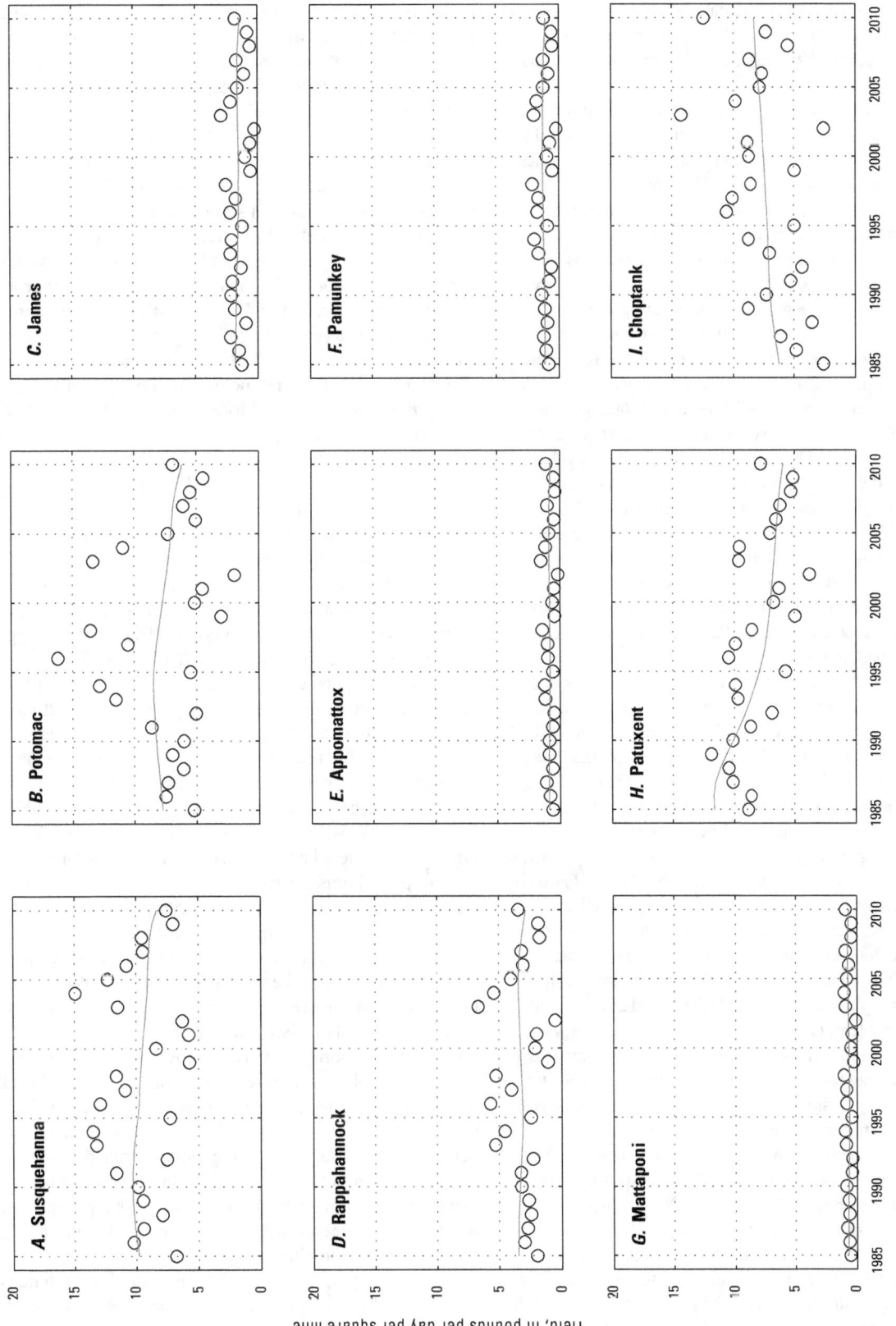

Figure 19. Nitrate yields (black circles) and flow-normalized yields (red line) for the nine RIM stations using the WRTDS method.

Phosphorus

Annual average and flow-normalized annual average total phosphorus yields for the nine RIM stations are presented in figure 20 for the period 1981 to 2010. Similar to the patterns in total nitrogen yields, total phosphorus yields from the Appomattox, Pamunkey, and Mattaponi Rivers stand out as having the smallest yields, generally less than 0.5 (lb/d)/mi^2. The Susquehanna River also stands out as having relatively low yields (fig. 20A) commonly less than 0.8 (lb/d)/mi^2 that are probably a direct result of total phosphorus being trapped in the reservoir just upstream from the Susquehanna RIM station. Unlike the predominance of RIM stations with improving conditions for total nitrogen and nitrate, total phosphorus at seven of the RIM stations exhibits positive (degrading) slopes during 1985 to 2010, which all steepen during 2001 to 2010 (fig. 20; table 5). The two RIM stations with the steepest positive slopes are the Rappahannock (4.0 percent per year, total increase of 99.8 percent) and the Pamunkey (2.4 percent per year; total increase of 60.5 percent) (table 5). Five RIM stations (Susquehanna, James, Appomattox, Mattaponi, and Choptank) have positive total phosphorus slopes that range from 0.0 to 0.6 percent per year (total increase of 1.3 and 16.1 percent, respectively) for the period 1985 to 2010. Two of the nine RIM stations that have negative slopes for 1985 to 2010 are the Patuxent (–2.4 percent per year, total reduction of –59.7 percent) and the Potomac (–0.5 percent per year, total reduction of –12.4) (table 5). The marked improvement at the point-source dominated Patuxent RIM station can be attributed, in part, to the phosphorus-based detergent ban that occurred around 1980 and improvements in the municipal wastewater system. The rate of total phosphorus delivery to the Chesapeake Bay increased dramatically at seven of the nine RIM stations for the period 2001 to 2010 (table 5). The greatest rates of change of total phosphorus yield for the period 2001 to 2010 occurred at the Rappahannock (6.9 percent per year), James (5.1 percent per year), Pamunkey (3.5 percent per year), Susquehanna (2.0 percent per year), Choptank (1.3 percent per year), and Appomattox (1.3 percent per year) Rivers (table 5). The Patuxent RIM station exhibits no measurable trend in total phosphorus yield for the period 2001 to 2010 compared to the considerable negative slope exhibited during 1985 to 2010. The Potomac RIM station (–0.6 percent per year) is the only one that has a negative slope for phosphorus yield for the period 2001 to 2010 (table 5). The long-term (1985 to 2010) trends in flow normalized total phosphorus yield indicate that the rate of total phosphorus delivery is increasing (degrading conditions) and this rate of total phosphorus delivery is further increasing at eight of the nine RIM stations for the more recent 2001 to 2010 period.

Annual average and flow-normalized annual average orthophosphorus yields for the nine RIM stations are presented in figure 21 for the period 1981 to 2010. Orthophosphorus is important because it is the form of phosphorus that is the most bioavailable; therefore, the mass of orthophosphorus delivered to the Chesapeake Bay is immediately available for biotic uptake and assimilation. From figure 21, four of the nine RIM stations (Potomac, James, Patuxent, and Choptank) stand out as having higher orthophosphorus yields routinely above

0.1 (lb/d)/mi^2. Eight of nine RIM stations have negative slopes for orthophosphorus for the period 1985 to 2010 (fig. 18) (table 6). These negative slopes range from –0.6 percent per year (total reduction of –14.2 percent) to –3.4 percent per year (total reduction of 86.2 percent). The Patuxent (–3.4 percent per year, total reduction of 84.6 percent) and the James (–3.4 percent per year, total reduction of 86.2 percent) exhibit the greatest reductions in orthophosphorus; these reductions can be attributed, in part, to the implementation of the phosphorus detergent ban in the early- to mid-1980s as well as improvements to municipal sewage treatment systems.

Three additional RIM stations that have sizable reductions in orthophosphorus are the Mattaponi (–2.0 percent per year, total reduction of 51.2 percent), Potomac (–1.9 percent per year, total reduction of 48.0 percent), and the Pamunkey (–1.5 percent per year; total reduction of 37.7 percent). The Choptank is the only RIM station that has a positive slope (1.8 percent per year, total increase of 44.2 percent) associated with orthophosphorus yield for the period 1985 to 2010. For the period 2001 to 2010, five of the nine RIM stations show an increased rate of orthophosphorus reduction compared to rates of reduction associated with the period 1985 to 2010; these stations are the James (–7.6 percent per year), Pamunkey (–5.1 percent per year), Mattaponi (–4.1 percent per year), Potomac (–3.9 percent per year) and Appomattox (–1.7 percent per year). The rates of orthophosphorus reduction have slowed at the Susquehanna, Rappahannock, and Patuxent RIM stations for the periods 2001 to 2010 compared to rates associated with 1985 to 2010 (table 6). The Choptank RIM station has greater rates of increasing orthophosphorus yields (3.6 percent per year) for the period 2001 to 2010 compared to the rate (1.8 percent per year) associated with 1985 to 2010. The results that have been presented related to changes in orthophosphorus yields over time show that efforts to control this constituent are working in eight of the nine watersheds. However, the results from the Choptank River, typically considered as a watershed with nitrogen-related and not phosphorus-related water-quality issues, show that the trends in total phosphorus and orthophosphorus yields are steadily increasing. This could be indicative of trends in other Eastern Shore tributaries, which have similar geology and land-use history. The results for flow-normalized trends in orthophosphorus are encouraging in that eight of nine RIM stations show orthophosphorus yields are improving over the long-term and short-term periods; however, these trend results for orthophosphorus are directly opposite of the trend results for total phosphorus flow-normalized yields where eight of the nine RIM stations show that the rate of total phosphorus delivery is increasing (becoming more degraded) during 2001 to 2010 compared to 1985 to 2010. One hypothesis for the dichotomy between orthophosphorus and total phosphorus is that the improvements in flow-normalized orthophosphorus yields are related to phosphorus controls associated with point sources in many of the watersheds (industrial and municipal wastewater discharges); whereas, increases in total phosphorus flow-normalized yields are related to increased loading from nonpoint sources (for example, phosphorus-saturated soils).

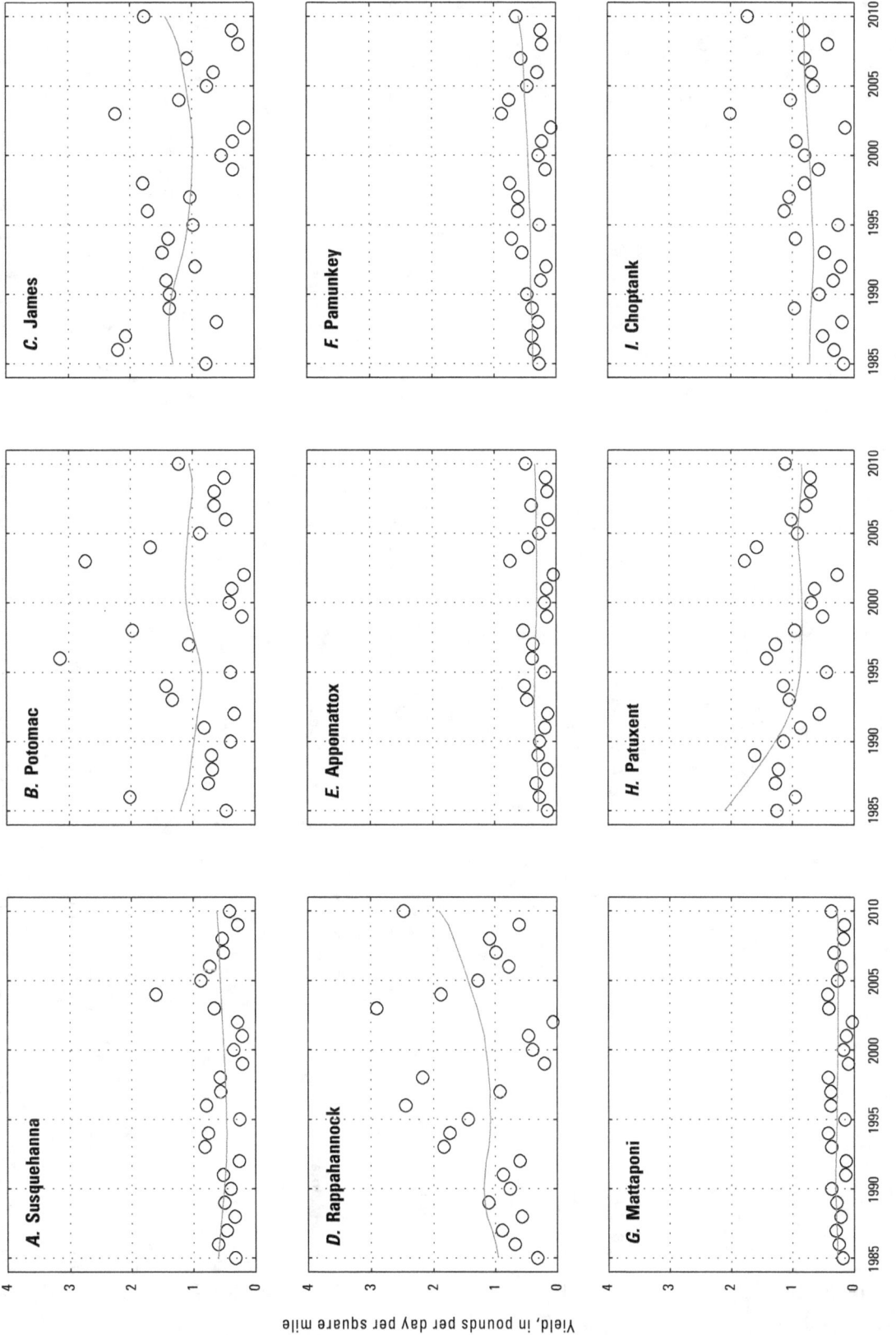

Figure 20.   Total phosphorus yields (black circles) and flow-normalized yields (red line) for the nine RIM stations using the WRTDS method.

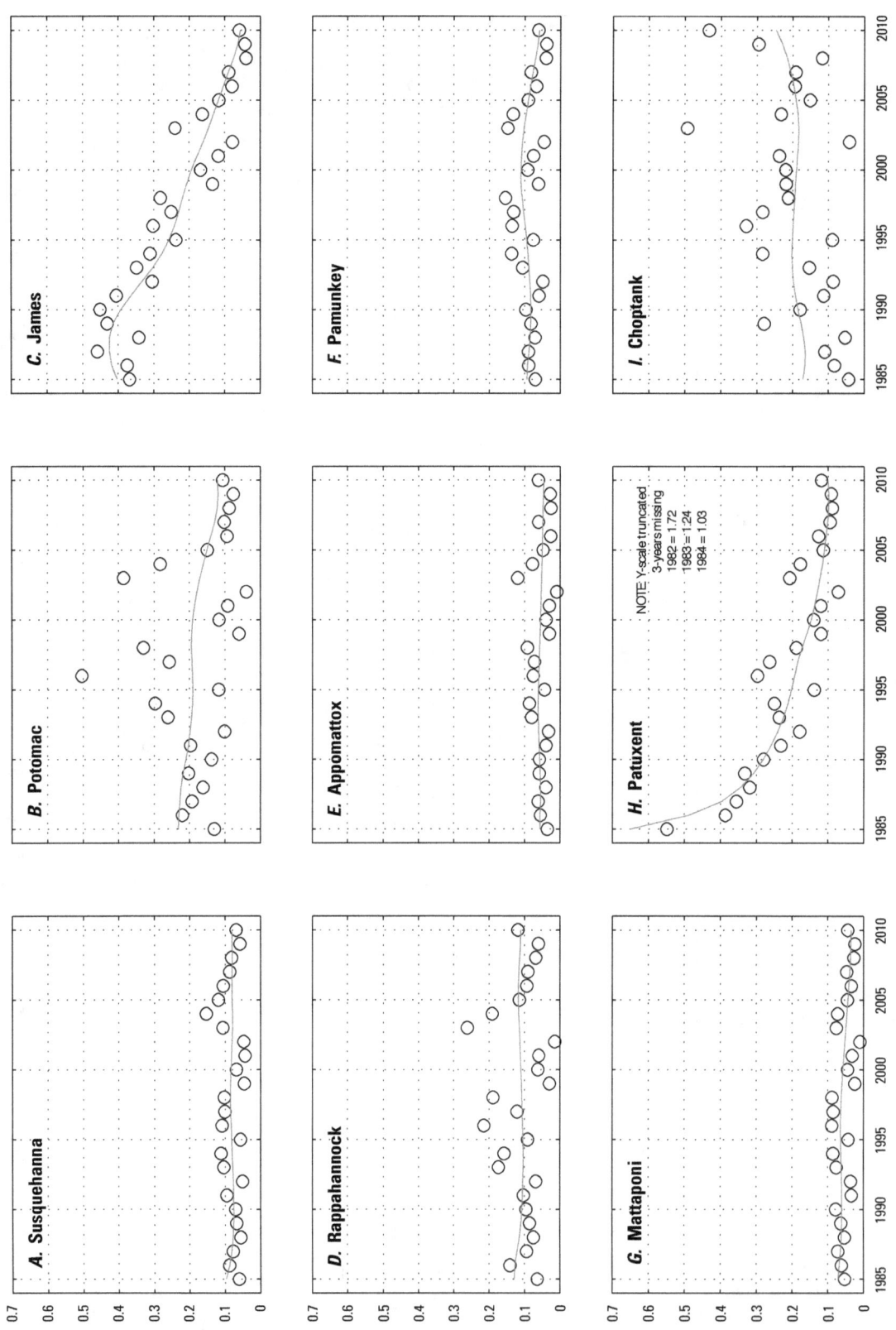

Figure 21. Orthophosphorus yields (black circles) and flow-normalized yields (red line) for the nine RIM stations using the WRTDS method.

Table 5. Total phosphorus trends in WRTDS flow-normalized yields and ESTIMATOR flow-adjusted concentration at the nine River Input Monitoring (RIM) stations for the time periods 1985 to 2010 and 2001 to 2010.

[%, percent; %/yr, percent per year; WRTDS flow-normalized yield results are presented as both the total percent change in yield and average slope (percent change in yield per year); ESTIMATOR changes in flow-adjusted concentration are reported as a 95-percent confidence interval (CI) for the average change in concentration per year; in the two ESTIMATOR columns of the table, shaded cells are those where the ESTIMATOR flow-adjusted concentration trend is significant, those with no shading are not significant. Pink shaded cells indicate that the ESTIMATOR flow-adjusted concentration trend and the WRTDS flow-normalized yield trend have the same sign; blue shaded cells indicate that the ESTIMATOR flow-adjusted concentration trend and the WRTDS flow-normalized yield trend have opposite signs. Black text indicates that the WRTDS trend in flow-normalized yield falls within the 95-percent confidence interval for ESTIMATOR flow-adjusted concentration; red text indicates that the WRTDS trend in flow-normalized yield falls outside the 95-percent confidence interval for ESTIMATOR flow-adjusted concentration; using this coding scheme, the most substantial apparent contradictions are those in the blue shaded cells (significant ESTIMATOR trend and opposite signs to the trend directions); all values are rounded to the nearest tenth]

| RIM station | WRTDS flow-normalized yield | | | | ESTIMATOR flow-adjusted concentration | |
| | 1985 to 2010 | | 2001 to 2010 | | 1985 to 2010 | 2001 to 2010 |
	Total change (%)	Slope (%/yr)	Total change (%)	Slope (%/yr)	Slope (Range of 95% CI, %/yr)	Slope (Range of 95% CI, %/yr)
Susquehanna	1.9	0.1	18.4	2.0	−0.8 to 0.2	−1.7 to 2.0
Potomac	−12.4	−0.5	−5.0	−0.6	−1.0 to 0.1	−6.0 to −3.4
James	9.4	0.4	46.0	5.1	−2.7 to −2.2	−7.6 to −5.1
Rappahannock	99.8	4.0	62.0	6.9	−1.1 to 0.3	−1.1 to 5.8
Appomattox	16.1	0.6	11.5	1.3	0.4 to 1.9	0.2 to 5.1
Pamunkey	60.5	2.4	31.2	3.5	2.8 to 5.3	−4.5 to −1.2
Mattaponi	1.2	0.0	3.6	0.4	−0.9 to 0.0	−2.1 to 0.7
Patuxent	−59.7	−2.4	0.2	0.0	−2.6 to −2.1	−4.3 to −0.8
Choptank	14.7	0.6	11.6	1.3	1.1 to 2.8	−2.1 to 1.2

Table 6. Orthophosphorus trends in WRTDS flow-normalized yields and ESTIMATOR flow-adjusted concentration at the nine River Input Monitoring (RIM) stations for the time periods 1985 to 2010 and 2001 to 2010.

[%, percent; %/yr, percent per year; WRTDS flow-normalized yield results are presented as both the total percent change in yield and average slope (percent change in yield per year); ESTIMATOR changes in flow-adjusted concentration are reported as a 95-percent confidence interval (CI) for the average change in concentration per year; in the two ESTIMATOR columns of the table, shaded cells are those where the ESTIMATOR flow-adjusted concentration trend is significant, those with no shading are not significant. Pink shaded cells indicate that the ESTIMATOR flow-adjusted concentration trend and the WRTDS flow-normalized yield trend have the same sign; blue shaded cells indicate that the ESTIMATOR flow-adjusted concentration trend and the WRTDS flow-normalized yield trend have opposite signs. Black text indicates that the WRTDS trend in flow-normalized yield falls within the 95-percent confidence interval for ESTIMATOR flow-adjusted concentration; red text indicates that the WRTDS trend in flow-normalized yield falls outside the 95-percent confidence interval for ESTIMATOR flow-adjusted concentration; using this coding scheme, the most substantial apparent contradictions are those in the blue shaded cells (significant ESTIMATOR trend and opposite signs to the trend directions); all values are rounded to the nearest tenth]

| RIM station | WRTDS flow-normalized yield | | | | ESTIMATOR flow-adjusted concentration | |
| | 1985 to 2010 | | 2001 to 2010 | | 1985 to 2010 | 2001 to 2010 |
	Total change (%)	Slope (%/yr)	Total change (%)	Slope (%/yr)	Slope (Range of 95% CI, %/yr)	Slope (Range of 95% CI, %/yr)
Susquehanna	−17.1	−0.7	−2.2	−0.2	0.6 to 3.6	−2.3 to 4.9
Potomac	−48.0	−1.9	−34.8	−3.9	−2.2 to −1.4	−7.4 to −4.1
James	−86.2	−3.4	−68.5	−7.6	−3.6 to −3.4	−10.0 to −9.2
Rappahannock	−14.2	−0.6	−0.4	0.0	−1.0 to 0.4	−2.9 to 2.4
Appomattox	−19.0	−0.8	−15.1	−1.7	−1.4 to −0.2	−1.5 to 3.6
Pamunkey	−37.7	−1.5	−46.3	−5.1	1.1 to 3.1	−8.3 to −7.0
Mattaponi	−51.2	−2.0	−36.9	−4.1	−2.1 to −1.4	−5.7 to −3.2
Patuxent	−84.6	−3.4	−27.6	−3.1	−3.2 to −3.0	−3.9 to −0.3
Choptank	44.2	1.8	32.5	3.6	0.5 to 2.6	3.3 to 9.8

Suspended Sediment

Annual average and flow-normalized annual average suspended sediment yields for the nine RIM stations are presented in figure 22 for the period 1981 to 2010 or for 2001 to 2010. The five RIM stations in Virginia (Rappahannock, Mattaponi, Pamunkey, James, and Appomattox) have suspended sediment monitoring records that span 2001 to 2010. Before 2001, water samples collected at these sites were analyzed for total suspended solids (TSS). Analysis for TSS (which uses an aliquot of the original sample) results in suspended-sediment concentrations that are biased low; the more reliable laboratory analysis is suspended-sediment concentration (SSC), which analyzes the entire water sample (Gray and others, 2000). In 2000, the SSC analysis was added to the analytical suite at the five RIM stations in Virginia. Based on 10 years of paired SSC and TSS results, the USGS RIM team decided to only use SSC derived suspended-sediment data for the determination of flux at these five RIM stations. At the Susquehanna, Patuxent, Potomac, and Choptank RIM stations, SSC has always been the primary analysis by which suspended-sediment concentration is determined.

Another feature that stands out in figure 22 is that four of the RIM stations have suspended sediment yields that are greater than 500 (lb/d)/mi^2 for about 50 percent of the years, (Potomac, James, Rappahannock, and Patuxent); four other RIM stations have suspended sediment yields below 500 (lb/d)/mi^2 virtually every year (Appomattox, Pamunkey, Mattaponi, and Choptank). The first group includes watersheds that drain extensive parts of the Piedmont Physiographic Province (fig. 1) and others areas of high topographic relief and (or) are areas of significant amounts of urban and agriculture land uses (Sprague and others, 2000). The latter group, except the Appomattox, almost entirely drains low-relief coastal plain areas and has very limited urban land use. The Appomattox River watershed is situated almost entirely within the Piedmont; however, Lake Chesdin, which is situated upstream from the Appomattox RIM station,

serves as a sink for suspended sediment. The Susquehanna (fig. 22A) is unique because it exhibits two different sediment transport processes such that (1) during average discharge years, such as 2010, the reservoir upstream from the RIM station acts as a sink for suspended sediment, thus reducing the average annual yield, and in most of those years the yield is below 500 (lb/d)/mi^2 and (2) during years with extreme high-flow events, such as 2004, the reservoir upstream from the RIM station acts as a source of suspended sediment and can greatly exceed this level (reaching a maximum of over 2,500 mg/L in 2004 (Langland, 2009). From the standpoint of landscape features, the Susquehanna is much more like the Potomac and James Rivers, for example, but the presence of the reservoir gives it a very different pattern of variability. The extent to which annual yields have changed outside the year-to year variations in discharge is provided in table 7 for two time periods, 1985 to 2010 and 2001 to 2010. For the period 1985 to 2010, the two largest watersheds, Susquehanna and Potomac, have positive slopes associated with the flow-normalized yields of 3.5 percent per year (total increase of 86.5 percent) and 0.5 percent per year (total increase of 12.2 percent), respectively (table 7). At the Patuxent and Choptank, suspended sediment yields have been decreasing at a rate of −0.8 and −0.9 percent per year, respectively (table 7). For the period 2001 to 2010, eight of the nine RIM stations exhibit patterns of increasing yields; although there is considerable variation, sediment yield is generally on the rise at these stations. Suspended sediment yield for the period 2001 to 2010 has the greatest rate of change at the Potomac and Susquehanna RIM stations where sediment yield is increasing at a rate of 9.9 and 7.9 percent per year, respectively (table 7). It is reasonable to think that this increase in the Potomac may be related to land-use changes in the watershed, while in the case of the Susquehanna River, the increase may be related to the decrease in the ability of the reservoir to trap sediment, and increased propensity for reservoir scour may be the dominant factor.

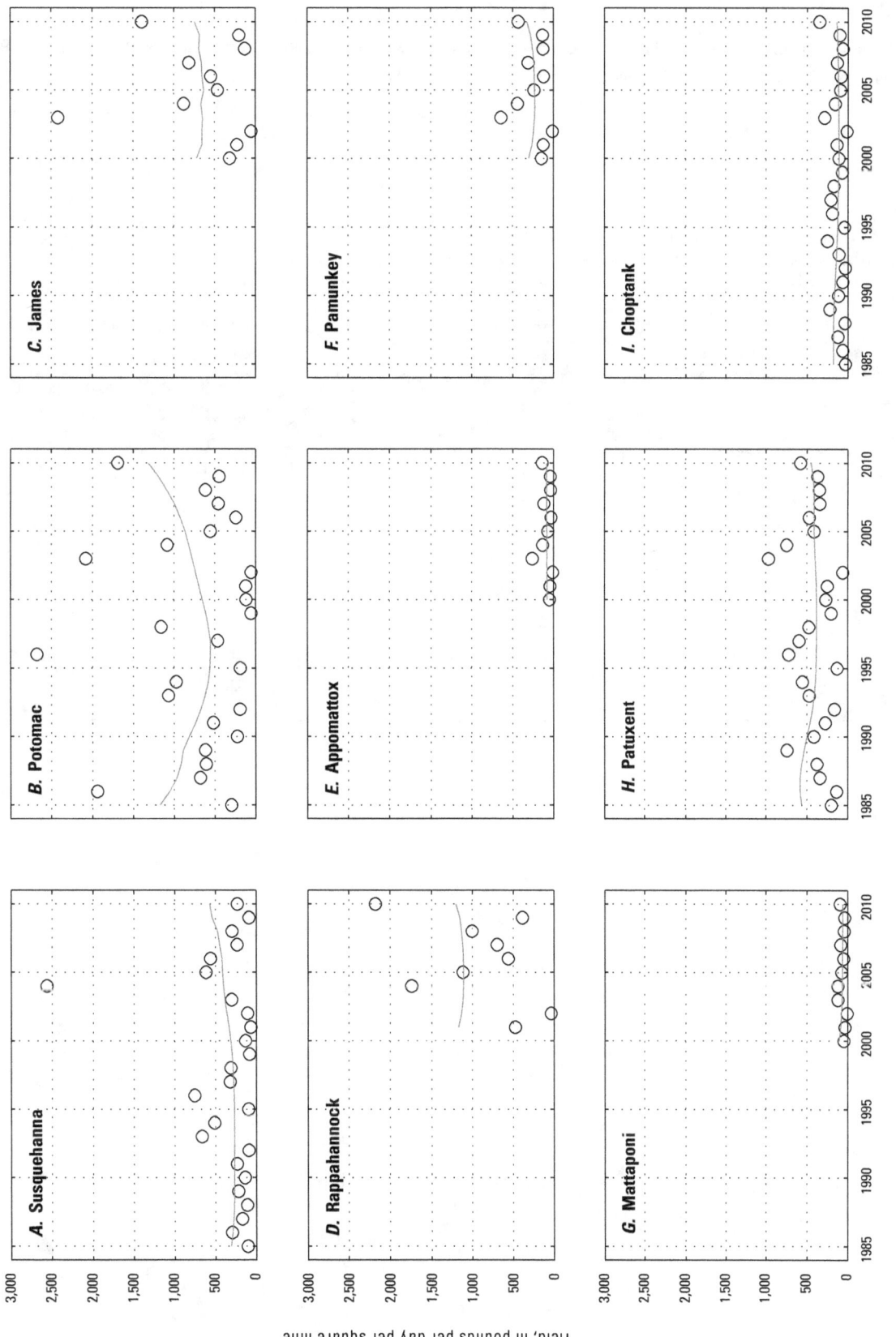

Figure 22. Suspended sediment yields (black circles) and flow-normalized yields (red line) for the nine RIM stations using the WRTDS method.

Table 7.   Suspended sediment trends in WRTDS flow-normalized yields and ESTIMATOR flow-adjusted concentration at the nine River Input Monitoring (RIM) stations for the time periods 1985 to 2010 and 2001 to 2010.

[%, percent; %/yr, percent per year; WRTDS flow-normalized yield results are presented as both the total percent change in yield and average slope (percent change in yield per year); ESTIMATOR changes in flow-adjusted concentration are reported as a 95-percent confidence interval (CI) for the average change in concentration per year; in the two ESTIMATOR columns of the table, shaded cells are those where the ESTIMATOR flow-adjusted concentration trend is significant, those with no shading are not significant. Pink shaded cells indicate that the ESTIMATOR flow-adjusted concentration trend and the WRTDS flow-normalized yield trend have the same sign; blue shaded cells indicate that the ESTIMATOR flow-adjusted concentration trend and the WRTDS flow-normalized yield trend have opposite signs. Black text indicates that the WRTDS trend in flow-normalized yield falls within the 95-percent confidence interval for ESTIMATOR flow-adjusted concentration; red text indicates that the WRTDS trend in flow-normalized yield falls outside the 95-percent confidence interval for ESTIMATOR flow-adjusted concentration; using this coding scheme, the most substantial apparent contradictions are those in the blue shaded cells (significant ESTIMATOR trend and opposite signs to the trend directions); all values are rounded to the nearest tenth]

RIM station	WRTDS flow-normalized yield				ESTIMATOR flow-adjusted concentration	
	1985 to 2010		2001 to 2010		1985 to 2010	2001 to 2010
	Total change (%)	Slope (%/yr)	Total change (%)	Slope (%/yr)	Slope (Range of 95% CI, %/yr)	Slope (Range of 95% CI, %/yr)
Susquehanna	86.5	3.5	71.1	7.9	−1.1 to −0.3	1.5 to 9.2
Potomac	12.2	0.5	89.1	9.9	−2.8 to −2.0	−4.4 to 4.0
James	NA	NA	14.3	1.6	NA	−3.0 to 5.4
Rappahannock	NA	NA	2.4	0.3	NA	−3.3 to 6.3
Appomattox	NA	NA	10.1	1.1	NA	−1.4 to 5.0
Pamunkey	NA	NA	23.1	2.6	NA	−0.6 to 12.6
Mattaponi	NA	NA	−28.7	−3.2	NA	−6.0 to −1.3
Patuxent	−20.8	−0.8	16.1	1.8	−2.1 to −1.4	−2.1 to 7.1
Choptank	−22.5	−0.9	26.0	2.9	−1.8 to −0.8	2.0 to 12.7

Summary and Conclusions

The accurate determination of nutrient (nitrogen and phosphorus) and sediment fluxes and changes in fluxes over time are essential components that can be used by water-resource managers to manage and improve the structure and function of the Chesapeake Bay ecosystem. The U.S. Geological Survey (USGS) has been responsible for providing annual nutrient and sediment flux data and trends that describe the extent to which water-quality conditions are changing within the major Chesapeake Bay tributaries. For the investigation summarized herein, the USGS compared two multiple linear regression approaches for the determination of annual fluxes and changes in annual fluxes over time. The two regression approaches compared are the traditionally used ESTIMATOR and the newly developed Weighted Regression on Time, Discharge, and Season (WRTDS). The overall objective for this investigation was to evaluate each model's performance in reproducing observed nutrient and sediment concentrations and fluxes and to determine which model ultimately ensures the highest level of accuracy in annual flux estimates provided to the Chesapeake Bay Program. The model comparison focused on three questions: (1) What are the differences between each model's functional form and how each model is constructed? (2) Which model produces discrete estimates of flux with the greatest accuracy and least amount of bias? and (3) How different would the historical estimates of annual flux have been had WRTDS been used instead of ESTIMATOR? One additional point of comparison was an evaluation of each model's ability to determine the changes in annual flux once the year-to-year variations in discharge have been accounted for. These results are displayed for total nitrogen, nitrate, total phosphorus, orthophosphorus, and suspended-sediment concentration data collected at the nine USGS operated River Input Monitoring (RIM) stations located on the Susquehanna, Potomac, James, Rappahannock, Appomattox, Pamunkey, Mattaponi, Patuxent, and Choptank Rivers.

The first comparison addressed the differences between each model's functional form and construction. Both ESTIMATOR and WRTDS predict water-quality constituent concentrations on the basis of river discharge, time, and season. ESTIMATOR is composed of seven parameters (discharge, discharge-squared, time, time-squared, two seasonal terms, and a model constant); whereas WRTDS is composed of five parameters (discharge, time, two seasonal terms, and a model constant). The functional form associated with ESTIMATOR allows constituent concentration to vary linearly and quadratically over the full range of discharge and time. The WRTDS model is much more flexible in terms of functional form even though the fitting process for any given set of explanatory variables constrains these relations to be (locally) linear. The greatest control governing the flexibility of each model, however, is directly related to how each model is constructed. ESTIMATOR models are constructed for a series of 9-year moving windows that estimate concentration and flux for the middle (5th) year and step forward in time

1 year at a time. New ESTIMATOR model coefficients are established for every 9-year window using all observed data for that window. Conversely, WRTDS models are constructed for each unique combination of discharge and time, and only observed data collected during similar discharges and times (determined by weighting water-quality observations based on distance of observation from modeled condition with respect to time, season, and discharge) are used to define the model coefficients. The use of model windows enable both models to account for changing relations between constituent concentration and flow and time better than if a single model was used for the entire period. The 9-year windows used for determining ESTIMATOR model coefficients is shown to have less flexibility relative to WRTDS. For example, ESTIMATOR accurately reproduces observed constituent concentration when the concentration-discharge relation is expressed as a linear or quadratic function (that is, zero or one point of inflection) or when deviations from a linear/quadratic form are well accounted for by other terms in the model such as time or season; otherwise, ESTIMATOR is unable to fully reproduce concentration-discharge relations that are more sinuous in form (that is, exhibit two points of inflection). WRTDS, on the other hand, has much greater flexibility because it only reproduces a single concentration for a given unique combination of discharge and time and only uses similar observations to define the model coefficients. Therefore, WRTDS has a greater ability, compared to ESTIMATOR, to reproduce complex observed concentration-discharge and (or) concentration-time relations.

The second model comparison addressed which model produced discrete estimates of concentration and flux with the greatest accuracy and least amount of bias. This comparison was accomplished by comparing discrete water-quality observations to associated estimates derived from ESTIMATOR and WRTDS. There are 45 combinations (nine RIM stations times five water-quality constituents) for comparison. With regard to model accuracy, WRTDS produced model estimates of flux for all 45 possible combinations that were more accurate than the estimates of flux that were derived from ESTIMATOR. The analysis of flux bias associated with WRTDS and ESTIMATOR-derived estimates resulted in the development of three categories of combinations. Category I contains combinations where no distinguishable differences existed between the modeled results with regard to flux bias. Of the 45 combinations, 28 combinations (62 percent) exhibited no or minor differences in flux bias between WRTDS and ESTIMATOR-derived estimates. Category II contains combinations where major differences with regard to flux bias occurred between the modeled results. Fifteen (33 percent) of the 45 combinations had at least a 10 percent difference in the value of flux bias associated with WRTDS and ESTIMATOR estimates. In all 15 combinations, WRTDS produced estimates of flux that were considerably less biased than ESTIMATOR-derived fluxes. The source of the elevated flux bias associated with ESTIMATOR resides in the complexity of the observed concentration-discharge relation. For all 15 combinations, the observed concentration-discharge relation exhibited a more sinuous pattern with two points of inflection at which the slope of the relation changed. ESTIMATOR can only reproduce concentration-discharge relations with either no points or one point of inflection (that is, the relation is either linear or quadratic in form); whereas, WRTDS is structured such that any number of shifts in the observed concentration-discharge relation can be reproduced. Category III contains two combinations (Susquehanna total phosphorus and Susquehanna suspended sediment) that cannot be categorized as either Category I or Category II. What makes these two combinations unique is that they are either particulate or particulate-dominated constituents collected just downstream from the Conowingo reservoir. For these two cases, ESTIMATOR outperformed WRTDS in reproducing concentrations and fluxes associated with the majority of the range of discharge; WRTDS outperforms ESTIMATOR in reproducing concentrations and fluxes associated with the highest discharges (greater than the 95 percentile). For Susquehanna suspended sediment, WRTDS produces flux estimates that are positively biased (flux-bias ratio equals 1.06) but are closer to 1.0 than the flux-bias ratio associated with ESTIMATOR, which is negatively biased (0.80). For Susquehanna total phosphorus, ESTIMATOR produces flux estimates that are negatively biased (flux-bias ratio equals 0.96) but are closer to 1.0 than flux estimates associated with WRTDS, which are positively biased (1.06). The difficulty WRTDS and ESTIMATOR have in reproducing the full range of the concentration-discharge relation, for Susquehanna suspended sediment and total phosphorus, is directly related to each model's functional form being inadequate in fully reproducing the transport processes (that is, shift from net depositional to net scour during extreme high discharges) governed by the Conowingo reservoir. Therefore, for the vast majority of combinations (44 of 45, 98 percent) WRTDS produces estimates of flux that are at worst equal and at best considerably less biased than estimates of flux derived from ESTIMATOR.

The third model comparison addressed how different the historical estimates of annual flux would be had WRTDS been used instead of ESTIMATOR. This comparison was made using the annual 1985 to 2010 flux estimates derived from ESTIMATOR and WRTDS. For total nitrogen, WRTDS would have generated annual fluxes that were, on average, −5.13 to 1.02 percent different than the annual fluxes generated by ESTIMATOR. The greatest difference (−5.13 percent) occurred at the Rappahannock RIM station. For nitrate, WRTDS would have generated annual fluxes that were, on average, −28.13 to −1.61 percent different. The James (−13.55 percent) and Rappahannock (−28.13 percent) are the only RIM stations to yield percent differences greater than 10 percent; these differences are related to the inability of ESTIMATOR to reproduce the complex concentration-discharge relation. For total phosphorus, WRTDS would have generated annual fluxes that were on average −17.68 to 12.38 percent different than the annual fluxes generated by ESTIMATOR. The Susquehanna (12.38 percent) and Rappahannock (−17.68 percent) are the only RIM stations to yield percent differences greater than

10 percent. The difference observed at the Susquehanna station is related to Category III discrepancies; while the difference observed at the Rappahannock is related to Category II discrepancies. For orthophosphorus, WRTDS would have generated annual fluxes that were, on average, −10.30 to 0.17 percent different than the annual fluxes generated by ESTIMATOR. The Patuxent (10.30 percent) is the only RIM station to yield a percent difference greater than 10 percent. This difference is related to Category II type discrepancies. For suspended sediment, WRTDS would have generated annual fluxes that were, on average, −39.50 to 38.33 percent different than the annual fluxes generated by ESTIMATOR. The Susquehanna (38.33 percent), Potomac (−19.34 percent), James (−19.20 percent), and Rappahannock (−39.50 percent) are the only RIM stations to yield percent differences greater than 10 percent. The difference in annual fluxes observed at the Susquehanna are directly related to Category III-type discrepancies; while differences at the Potomac, James, and Rappahannock RIM stations are all related to Category II-type discrepancies.

The last point of comparison is each model's ability to quantify the changes in annual flux once the year-to-year variations in discharge have been accounted for. ESTIMATOR is not able to quantify the changes in annual flux (as being unique from changes in concentration) once the year-to-year variations in flow are removed. ESTIMATOR has historically defined the changes in water-quality conditions as a trend in flow-adjusted concentration; however, this information is not completely indicative of how flux is changing over time because the functional form of ESTIMATOR assumes that the percentage changes in concentration for a given discharge and time of year are the same across all discharges and seasons. WRTDS allows a wide range of relations and thus is able to estimate flux changes that more closely match the kinds of changes actually occurring in the watershed. Conversely, WRTDS allows for the direct determination of changes in annual flux once year-to-year variations in discharge have been addressed (flow normalization). Temporal changes in flow-normalized flux can be attributed to changes in nutrient/sediment sources and (or) land uses and (or) implementation of water-quality improvement strategies. Temporal changes in flow-normalized flux (presented as changes in yield) were presented for two time periods, 1985 to 2010 and 2001 to 2010. Before the changes in WRTDS flow-normalized flux were presented, the WRTDS flow-normalized fluxes were compared to the flow-adjusted concentration trends produced by ESTIMATOR. There are 55 of 85 combinations (64 percent) where the estimated WRTDS flow-normalized trend falls within the 95-percent confidence interval associated with the ESTIMATOR flow-adjusted concentration trends (significant and non-significant trends). A total of 48 of the 85 combinations have ESTIMATOR flow-adjusted concentration trends that are significantly different from zero. Of these 48 combinations, there are 38 combinations (79 percent) where the trend in WRTDS flow-normalized flux is in complete agreement, with respect to the direction of change (that is, improving or degrading conditions), with

ESTIMATOR flow-adjusted concentration trends. There are only 10 of the 48 combinations (21 percent) where the direction of change differs between the trends in WRTDS flow-normalized flux and ESTIMATOR flow-adjusted concentration, and these differences for the 10 combinations are related to real differences between changes in concentrations and fluxes. Once the trends in WRTDS flow-normalized flux were compared to ESTIMATOR flow-adjusted concentration trends, the trends in WRTDS flow-normalized fluxes were further explored for total nitrogen, nitrate, total phosphorus, orthophosphorus, and suspended sediment at the nine RIM stations.

For total nitrogen, seven of nine RIM stations exhibited negative slopes (improving conditions, decreasing fluxes) in flow-normalized fluxes for the period 1985 to 2010; whereas two RIM stations, Choptank and Pamunkey, exhibited positive slopes (degrading condition, increasing fluxes). For the period 2001 to 2010, only four of the nine RIM stations exhibited negative slopes in flow-normalized fluxes, and of these four RIM stations, only the Rappahannock showed an increase in the rate of improvement compared to the rate during 1985 to 2010. The Patuxent exhibited the greatest rates of improvement with a rate of –2.0 percent per year (total improvement of 49.3 percent) during 1985 to 2010 and a rate of –1.2 percent per year during 2001 to 2010. Conversely, the Choptank exhibited the greatest rates of increase in flux with a rate of 0.3 percent per year (total degradation of 7.8 percent) during 1985 to 2010 and a rate of 0.8 percent per year during 2001 to 2010.

For nitrate, eight of nine RIM stations exhibited negative slopes in flow-normalized fluxes for the period 1985 to 2010; whereas only the Choptank exhibited a positive slope. For the period 2001 to 2010, eight of the nine RIM stations exhibited negative slopes in flow-normalized fluxes, and of these eight RIM stations, six stations exhibited increases in the rate of improvement. The Pamunkey RIM station exhibited the greatest rate of increase from –0.3 percent per year for 1985 to 2010 to –1.8 percent per year for 2001 to 2010. The Choptank is the only RIM station that exhibited positive slopes for nitrate flow-normalized fluxes with a rate of 1.3 percent per year during 1985 to 2010 and a rate of 1.1 percent per year during 2001 to 2010.

For total phosphorus, only two of nine RIM stations (Potomac and Patuxent) exhibited negative slopes in flow-normalized fluxes for the period 1985 to 2010. For the period 2001 to 2010, only one of the nine RIM stations (Potomac) exhibited negative slopes in flow-normalized fluxes. The rates of improvement in total phosphorus flow-normalized flux are –0.5 percent per year (total improvement of 13.0 percent) during 1985 to 2010 and –0.6 percent during 2001 to 2010. The Patuxent showed a considerable rate of improvement (–2.4 percent per year or total reduction of 59.7 percent) during 1985 to 2010; conversely, during 2001 to 2010 the Patuxent exhibited no detectable slope (0.0 percent per year). The Rappahannock had the greatest positive slope for both time periods with rates equal to 4.0 percent per year (total increase of 99.8 percent) during 1985 to 2001 and 6.9 percent per year during 2001 to 2010.

For orthophosphorus, eight of nine RIM stations exhibited negative slopes in flow-normalized fluxes for the period 1985 to 2010; whereas only the Choptank station exhibited a positive slope. For the period 2001 to 2010, eight of the nine RIM stations exhibited negative slopes in flow-normalized fluxes and of these eight RIM stations five exhibited increases in the rate of improvement. The James River exhibited the greatest change in the rate of improvement from –3.4 percent per year (total reduction of 86.2 percent) for 1985 to 2010 to –7.6 percent per year for 2001 to 2010. The Choptank is the only RIM station that exhibited positive slopes for orthophosphorus flow-normalized fluxes with a rate of 1.8 percent per year during 1985 to 2010 and a rate of 3.6 percent per year during 2001 to 2010.

For suspended sediment, eight of nine RIM stations exhibit positive slopes in flow-normalized fluxes for the period 2001 to 2010 with slopes ranging from 0.3 to 9.9 percent per year. The Potomac River exhibited the greatest rate of change from 0.5 percent per year during 1985 to 2010 to 9.9 percent per year during 2001 to 2010. The Mattaponi River is the only RIM station that had a negative slope for flow normalized flux during 2001 to 2010; the rate of decrease is –3.2 percent per year.

The results of this investigation have shown that WRTDS is a viable alternative to ESTIMATOR for the determination of annual nutrient and sediment fluxes at the nine Chesapeake Bay watershed RIM stations. WRTDS produces discrete concentrations and fluxes that are generally more accurate than those generated using ESTIMATOR, especially when the observed concentration-discharge relation is more sinuous in form with two points of inflection. The single most beneficial aspect associated with using WRTDS, however, is the added ability to report changes in flow-normalized annual fluxes that have occurred as a result of human activities in the watershed. Information on the changes in flow-normalized flux improves the relevancy of the USGS RIM data to the Chesapeake Bay Program and allows water-resource managers to better understand the flux of nutrients and sediment to the Chesapeake Bay and to directly measure the efficacy of their efforts to limit the delivery of nutrients and sediment to the bay.

Acknowledgments

The authors wish to thank the Virginia Department of Environmental Quality, Maryland Department of Natural Resources, and U.S. Environmental Protection Agency Chesapeake Bay Program whose continued support and collaboration made this report possible. The authors also thank Laura Medalie and Jeff Chanat of the USGS for providing thorough technical reviews. Many other individuals within the USGS and Chesapeake Bay Program contributed to the report through various discussions and input; their contributions are sincerely appreciated.

References Cited

Aulenbach, B.T., Buxton, H.T., Battaglin, W.A., and Coupe, R.H., 2007, Streamflow and nutrient fluxes of the Mississippi-Atchafalaya River basin and subbasins for the period of record through 2005: U.S. Geological Survey Open-File Report 2007–1080, accessed September 17, 2012, at *http://toxics.usgs.gov/pubs/of-2007-1080/index.html*.

Box, J.B., and Mossa, JoAnn, 1999, Sediment, land use, and freshwater mussels—Prospects and problems: Journal of the North American Benthological Society, v. 18, no. 1, p. 99–117.

Cairns, J., Jr., 1977, Aquatic ecosystem assimilative capacity: Fisheries, v. 2, no. 2, p. 5–7.

Chesapeake Bay Foundation, 2010, State of the Bay 2010: Accessed June 25, 2012, at *http://www.cbf.org/document. doc?id=596*.

Chesapeake Bay Program, 2000, CHESAPEAKE 2000: Accessed May 30, 2012, at *http://www.chesapeakebay.net/ documents/cbp_12081.pdf*.

Cohn, T.A., 2005, Estimating contaminant loads in rivers: An application of adjusted maximum Likelihood to type 1 censored data: Water Resources Research, v. 41, W07003, doi:10.1029/2004WR003833.

Cohn, T.A., Caulder, D.L., Gilroy, E.J., Zynjuk, L.D., and Summers, R.M., 1992, The validity of a simple log-linear model for estimating fluvial constituent loads—An empirical study involving nutrient loads entering Chesapeake Bay: Water Resources Research, v. 28, no. 9, p. 2353–2364.

Cohn, T.A., DeLong, L.L., Gilroy, E.J., Hirsch, R.M., and Wells, R.M., 1989, Estimating constituent loads: Water Resources Research, v. 25, no. 5, p. 937–942.

Cooper, S.R., and Brush, G.S., 1991, Long-term history of Chesapeake Bay anoxia: Science, v. 254, p. 992–996.

Dennison, W.C., Orth, R.J., Moore, K.A., Stevenson, J.C., Carter, Virginia, Kollar, Stan, Bergstrom, P.W., and Batiuk, R.A., 1993, Assessing water quality with submersed aquatic vegetation: BioScience, v. 43, no. 2, p. 86–94.

Duan, Naihua, 1983, Smearing estimate—A nonparametric retransformation method: Journal of the American Statistical Association, v. 78, p. 605–610.

Gilroy, E.J., Hirsch, R.M., Kirby, W.H., and Cohn, T.A., 1990, Mean square error of regression-based constituent transport estimates: Water Resources Research, v. 26, no. 9, p. 2069–2077.

Gray, J.R., Glysson, G.D., Turcios, L.M., Schwarz, G.E., 2000, Comparability of suspended-sediment concentration and total suspended solids data: U.S. Geological Survey Water-Resources Investigations Report 00–4191, 20 p.

Helsel, D.R., 2012, Statistics for censored environmental data using Minitab and R, (2d ed.): Hoboken, N.J., John Wiley & Sons, 324 p.

Hirsch, R.M., Moyer, D.L., and Archfield, S.A., 2010, Weighted regression on time, discharge, and season (WRTDS), with an application to Chesapeake Bay river inputs: Journal of American Water Resources Association, v. 46, no. 5, p. 857–1064.

Judge, G.G., Griffiths, W.E., Hill, R.C., Lutkepohl, K., and Lee, T.C., 1985, Qualitative and limited dependent variable models: Chapter 18, *in* The Theory and Practice of Econometrics: New York, John Wiley & Sons, 1,019 p.

Langland, M.J., 2009, Bathymetry and sediment-storage capacity change in three reservoirs on the Lower Susquehanna River, 1996–2008: U.S. Geological Survey Scientific Investigations Report 2009–5110, 21 p.

Langland, M.J., Lietman, P.L., and Hoffman, S.A., 1995, Synthesis of nutrient and sediment data for watersheds within the Chesapeake Bay drainage basin: U.S. Geological Survey Water-Resources Investigations Report 95–4233, 121 p.

Langland, M.J., Raffensperger, J.P., Moyer, D.L., Landwehr, J.M., and Schwarz, G.E., 2006, Changes in streamflow and water quality in selected nontidal basins in the Chesapeake Bay watershed, 1985–2004: U.S. Geological Survey Scientific Investigations Report 2006–5178, 75 p.

Lenat, D.R., Penrose, D.L., and Eagleson, K.W., 1981, Variable effects of sediment addition on stream benthos: Hydrobiologia, v. 79, no. 2, p. 187–194.

Madsen, J.D., Chambers, P.A., James, W.F., Koch, E.W., and Westlake, D.F., 2001, The interaction between water movement, sediment dynamics, and submersed macrophytes: Hydrobiologia, v. 444, nos. 1–3, p. 61–84.

Medalie, L., Hirsch, R.M., and Archfield, S.A., 2012, Use of flow-normalization to evaluate nutrient concentration and flux changes in Lake Champlain tributaries, 1990–2009: Journal of Great Lakes Research, v. 38, supplement 1, p. 58–67.

Milly, P.C.D., Betancourt, J., Falkenmark, M., Hirsch, R.M., Kundzewicz, Z.W., Lettenmaier, D.P., and Stouffer, R.J., 2008, Stationarity is dead: Whither water management?: Science, v. 319, no. 5863, p. 573–574.

Nixon, S.W., 1987, Chesapeake Bay nutrient budgets—A reassessment: Biogeochemistry, v. 4, p. 77–90.

Officer, C.B., Biggs, R.B., Talf, J.L., Cronin, L.E., Tyler, M.A., and Boynton, W.R., 1984 Chesapeake Bay anoxia—Origin, development, and significance: Science, v. 223, p. 22–27.

Schlesinger, W.H., 1997, Biogeochemistry—An analysis of global change: New York, Academic Press, 588 p.

Sprague, L.A., Hirsch, R.M., and Aulenbach, B.T., 2011, Nitrate in the Mississippi River and its tributaries, 1980 to 2008: Are we making progress?: Environmental Science & Technology, v. 45, no. 17, p. 7209–7216.

Sprague, L.A., Langland, M.J., Yochum, S.E., Edwards, R.E., Blomquist, J.D., Phillips, S.W., Shenk, G.W., and Preston, S.D., 2000, Factors affecting nutrient trends in major rivers of the Chesapeake Bay watershed: U.S. Geological Survey Water-Resources Investigations Report 00–4218, 98 p.

Tobin, James, 1958, Estimation relationships for limited dependent variables: Econometrica, v. 26, no. 1, p. 24–36.

Tukey, J.W., 1977, Exploratory data analysis: Reading, Mass., Addison-Wesley, 688 p.

U.S. Environmental Protection Agency, 1983, Chesapeake Bay—A framework for action: Philadelphia, Pa., U.S. Environmental Protection Agency, Region 3, 186 p.

U.S. Environmental Protection Agency, 2010, Chesapeake Bay TMDL: Accessed June 25, 2012, at *http://www.epa.gov/chesapeakebaytmdl/*.

Waters, T.F., 1995, Sediment in streams: Sources, biological effects, and control: Bethesda, Md., American Fisheries Society, 251 p.

Yochum, S.E., 2000, A revised load estimation procedure for the Susquehanna, Potomac, Patuxent, and Choptank Rivers: U.S. Geological Survey Water-Resources Investigations Report 00–4156, 49 p.

Appendix 1. Total nitrogen, nitrate, total phosphorus, orthophosphorus, and suspended sediment collected at the nine River Input Monitoring stations in the Chesapeake Bay watershed showing *(A)* observed concentration (red dots) versus discharge relation, *(B)* observed (red dots) and ESTIMATOR-predicted (black dots) concentration versus discharge relation, *(C)* residual (observed minus predicted) plot for ESTIMATOR predictions, *(D)* observed (red dots) and WRTDS-predicted (black dots) concentration versus discharge relation, and *(E)* residual (observed minus predicted) plot for WRTDS predictions.

_____

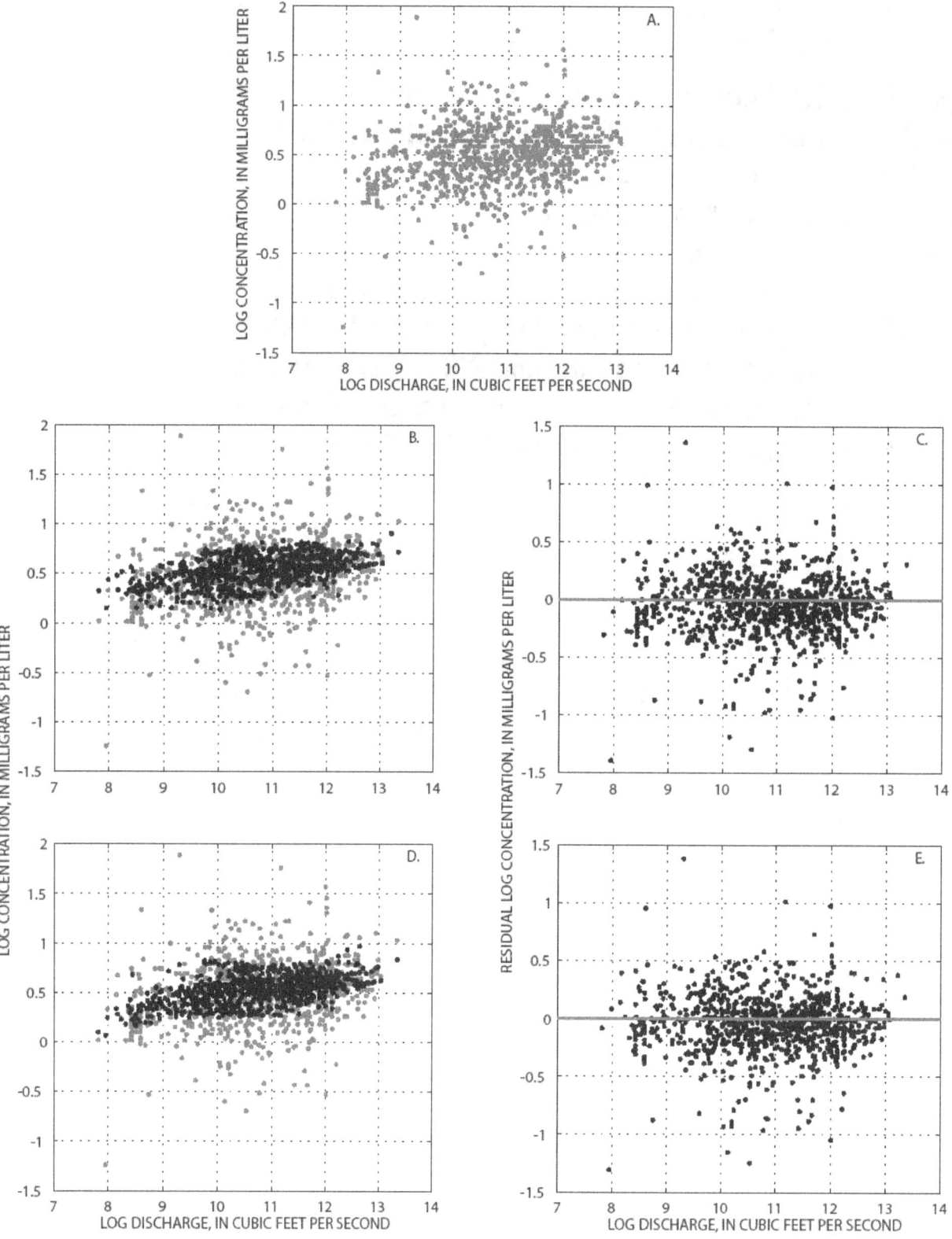

Figure 1–1.   Total nitrogen at Susquehanna River at Conowingo, Maryland (USGS Station ID 01578310), showing the *(A)* observed concentration (red dots) versus discharge relation, *(B)* observed (red dots) and ESTIMATOR-predicted (black dots) concentration versus discharge relation, *(C)* residual (observed minus predicted) plot for ESTIMATOR predictions, *(D)* observed (red dots) and WRTDS-predicted (black dots) concentration versus discharge relation, and *(E)* residual (observed minus predicted) plot for WRTDS predictions.

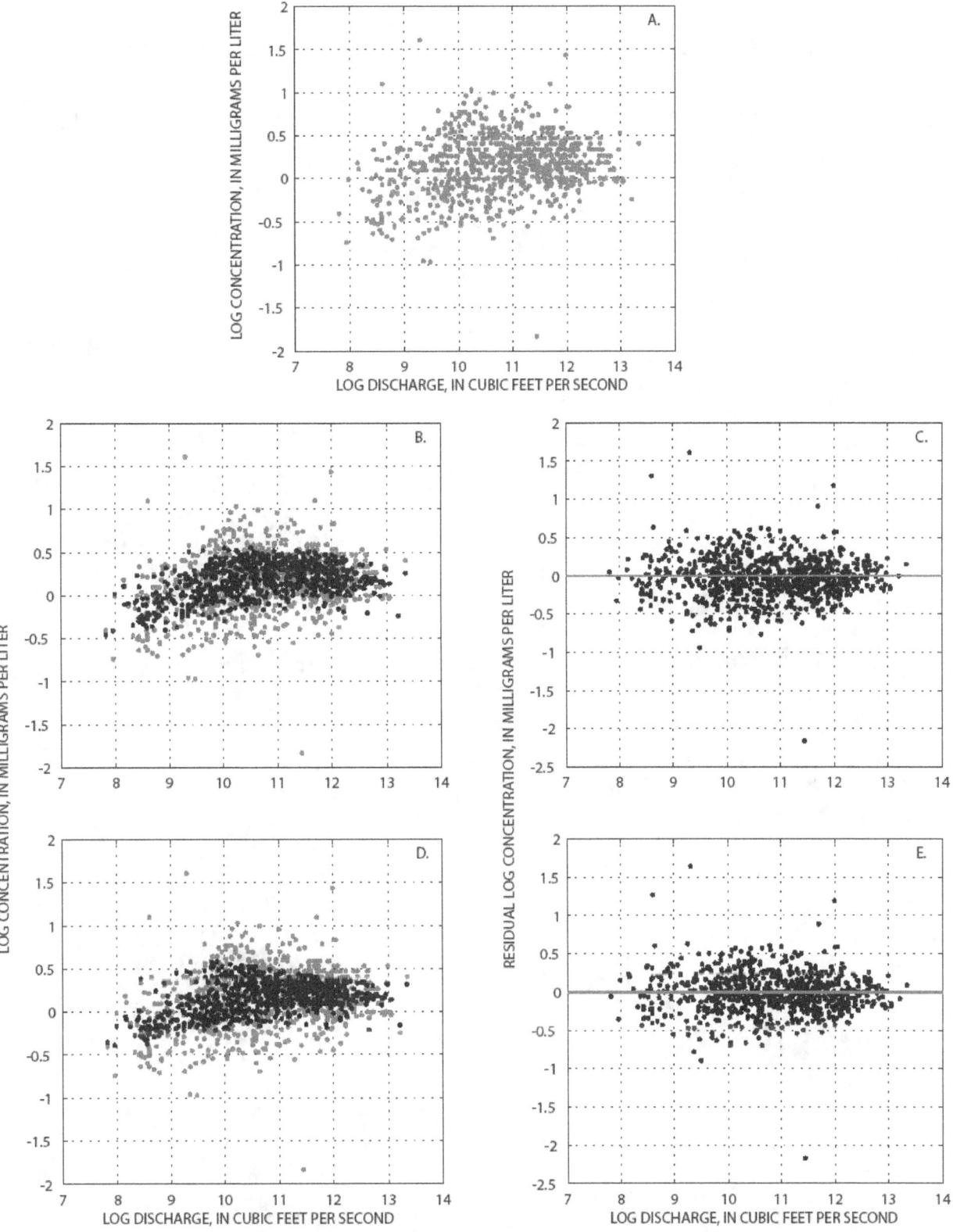

Figure 1–2.   Nitrate at Susquehanna River at Conowingo, Maryland (USGS Station ID 01578310), showing the *(A)* observed concentration (red dots) versus discharge relation, *(B)* observed (red dots) and ESTIMATOR-predicted (black dots) concentration versus discharge relation, *(C)* residual (observed minus predicted) plot for ESTIMATOR predictions, *(D)* observed (red dots) and WRTDS-predicted (black dots) concentration versus discharge relation, and *(E)* residual (observed minus predicted) plot for WRTDS predictions.

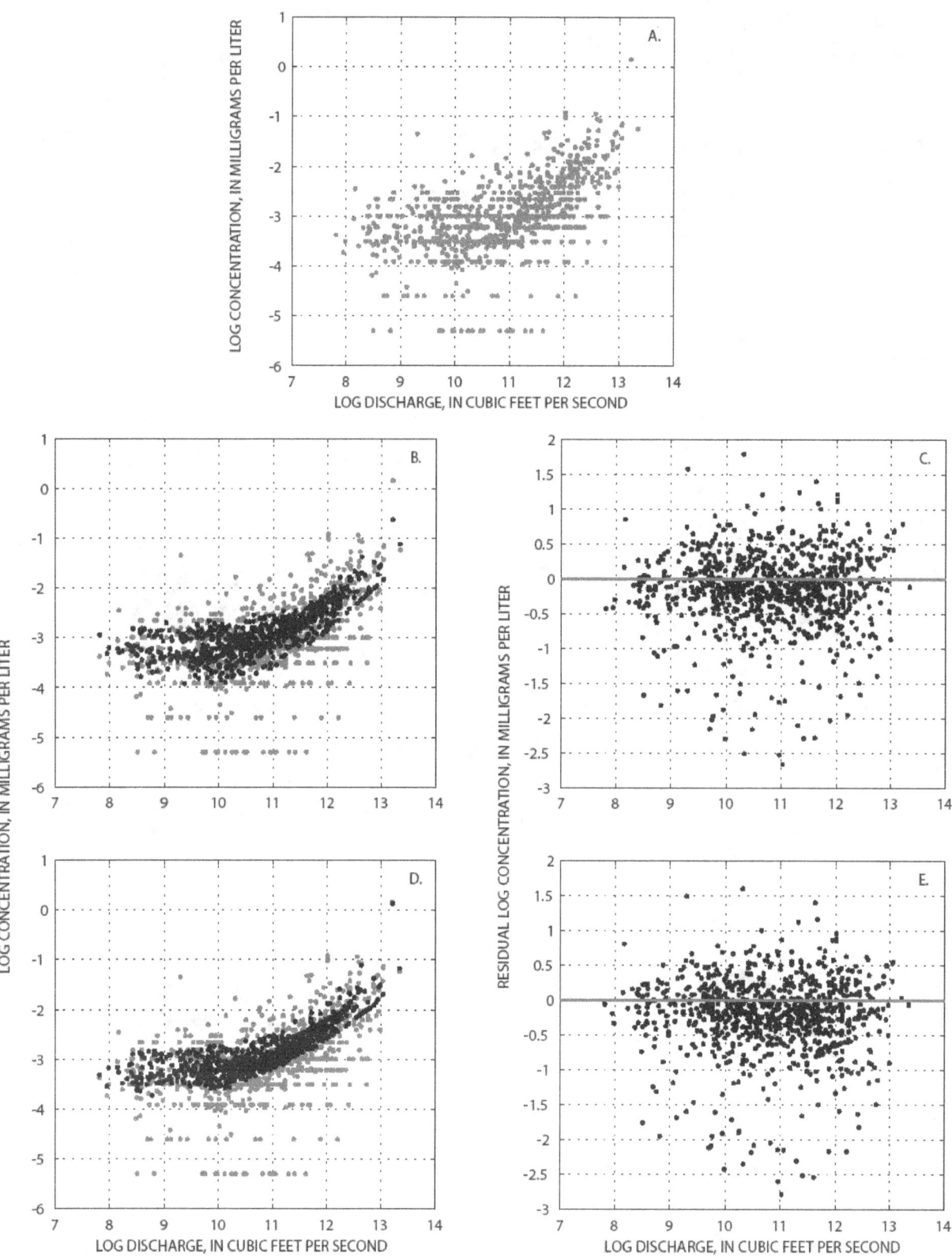

Figure 1–3. Total phosphorus at Susquehanna River at Conowingo, Maryland (USGS Station ID 01578310), showing the (A) observed concentration (red dots) versus discharge relation, (B) observed (red dots) and ESTIMATOR-predicted (black dots) concentration versus discharge relation, (C) residual (observed minus predicted) plot for ESTIMATOR predictions, (D) observed (red dots) and WRTDS-predicted (black dots) concentration versus discharge relation, and (E) residual (observed minus predicted) plot for WRTDS predictions.

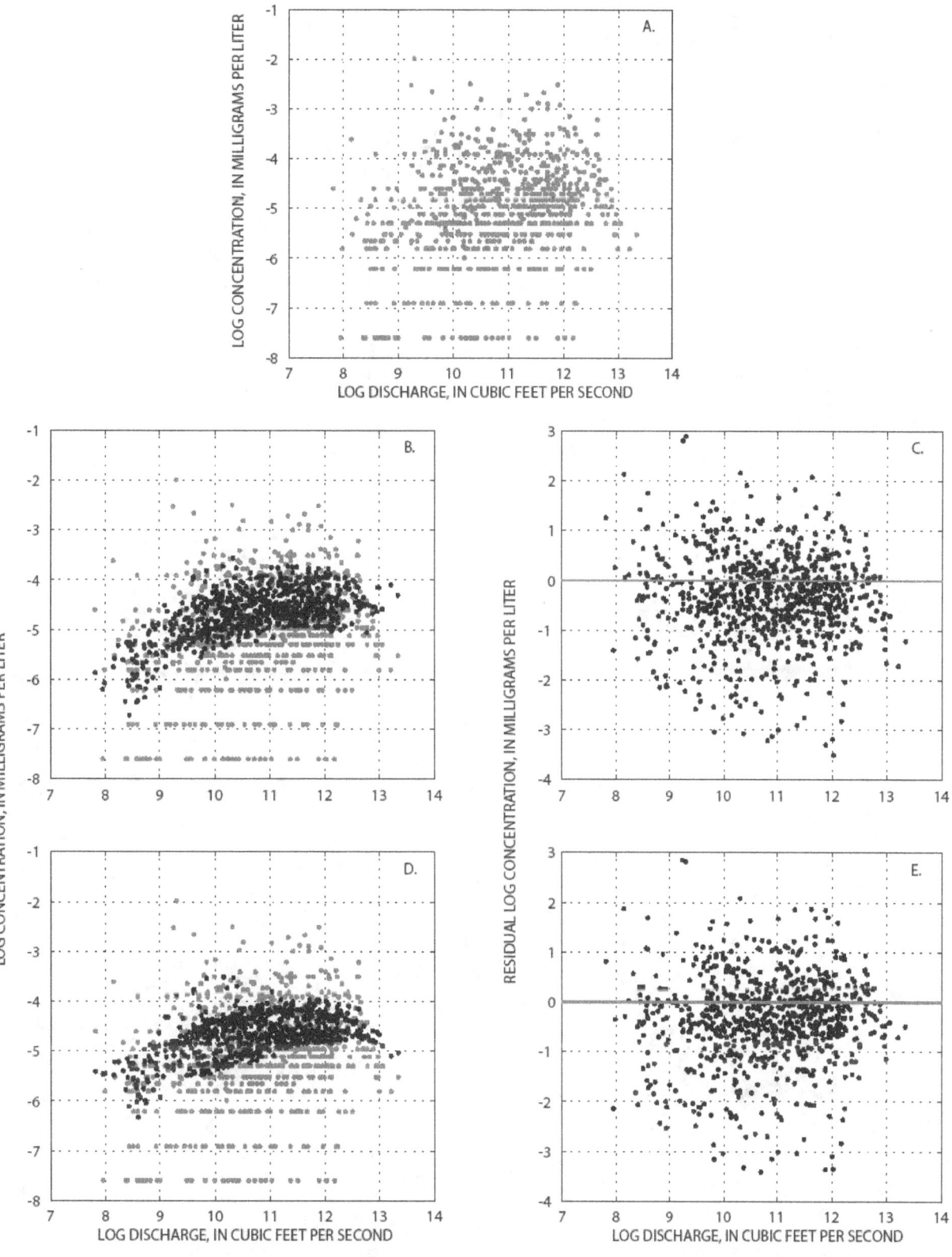

Figure 1–4.   Orthophosphorus at Susquehanna River at Conowingo, Maryland (USGS Station ID 01578310), showing the *(A)* observed concentration (red dots) versus discharge relation, *(B)* observed (red dots) and ESTIMATOR-predicted (black dots) concentration versus discharge relation, *(C)* residual (observed minus predicted) plot for ESTIMATOR predictions, *(D)* observed (red dots) and WRTDS-predicted (black dots) concentration versus discharge relation, and *(E)* residual (observed minus predicted) plot for WRTDS predictions.

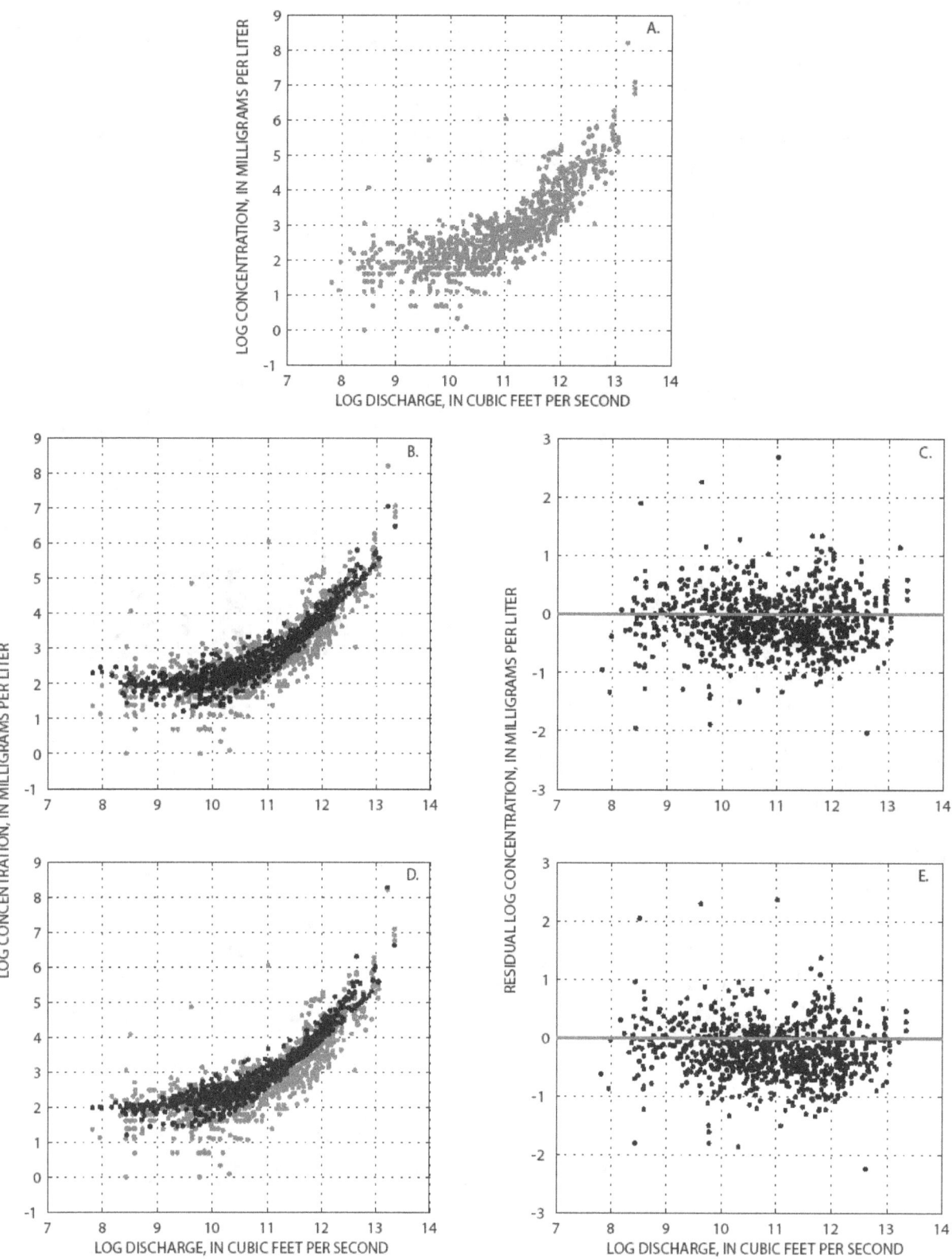

Figure 1–5.   Suspended sediment at Susquehanna River at Conowingo, Maryland (USGS Station ID 01578310), showing the *(A)* observed concentration (red dots) versus discharge relation, *(B)* observed (red dots) and ESTIMATOR-predicted (black dots) concentration versus discharge relation, *(C)* residual (observed minus predicted) plot for ESTIMATOR predictions, *(D)* observed (red dots) and WRTDS-predicted (black dots) concentration versus discharge relation, and *(E)* residual (observed minus predicted) plot for WRTDS predictions.

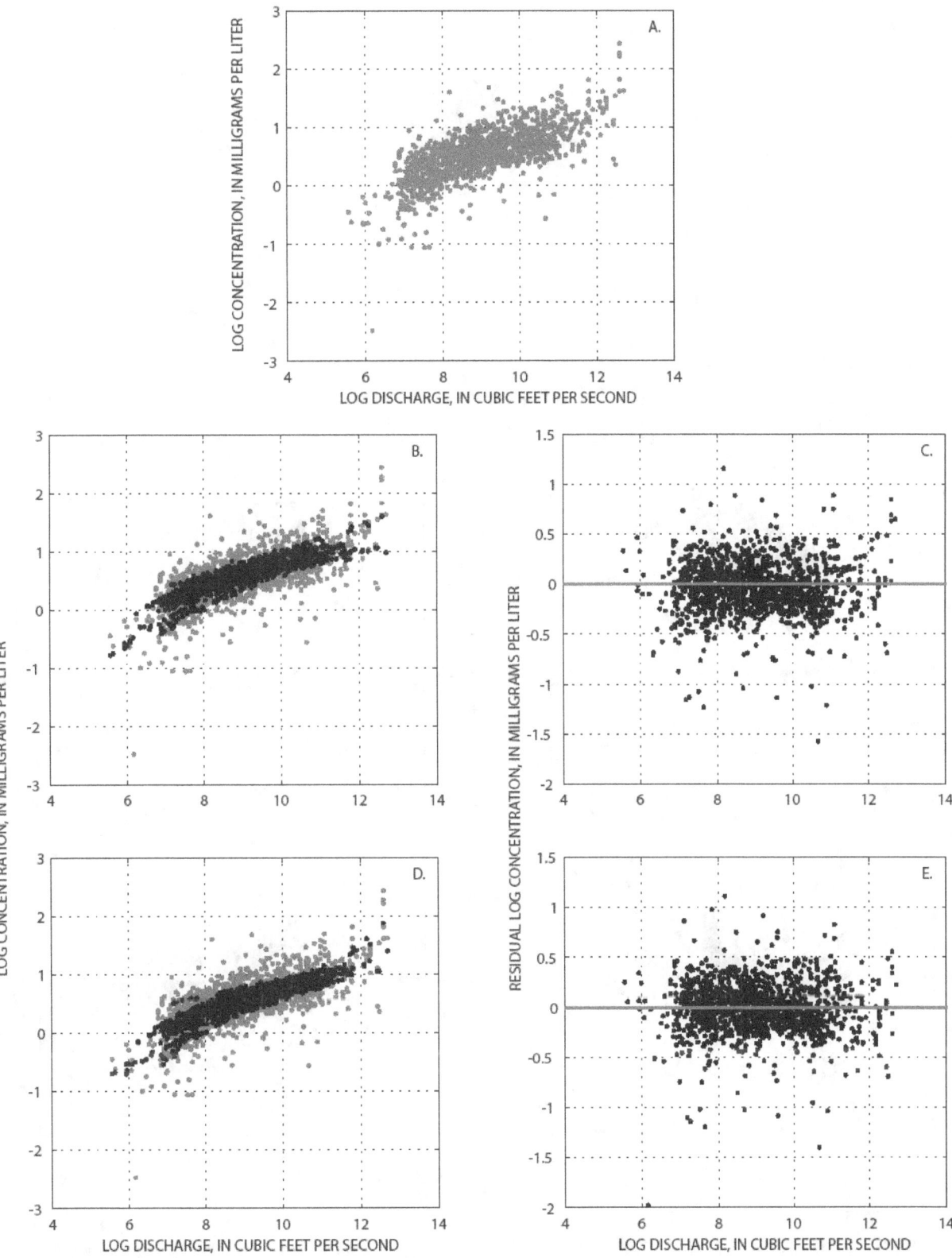

Figure 1–6. Total nitrogen at Potomac River at Chain Bridge at Washington, D.C. (USGS Station ID 01646580), showing the *(A)* observed concentration (red dots) versus discharge relation, *(B)* observed (red dots) and ESTIMATOR-predicted (black dots) concentration versus discharge relation, *(C)* residual (observed minus predicted) plot for ESTIMATOR predictions, *(D)* observed (red dots) and WRTDS-predicted (black dots) concentration versus discharge relation, and *(E)* residual (observed minus predicted) plot for WRTDS predictions.

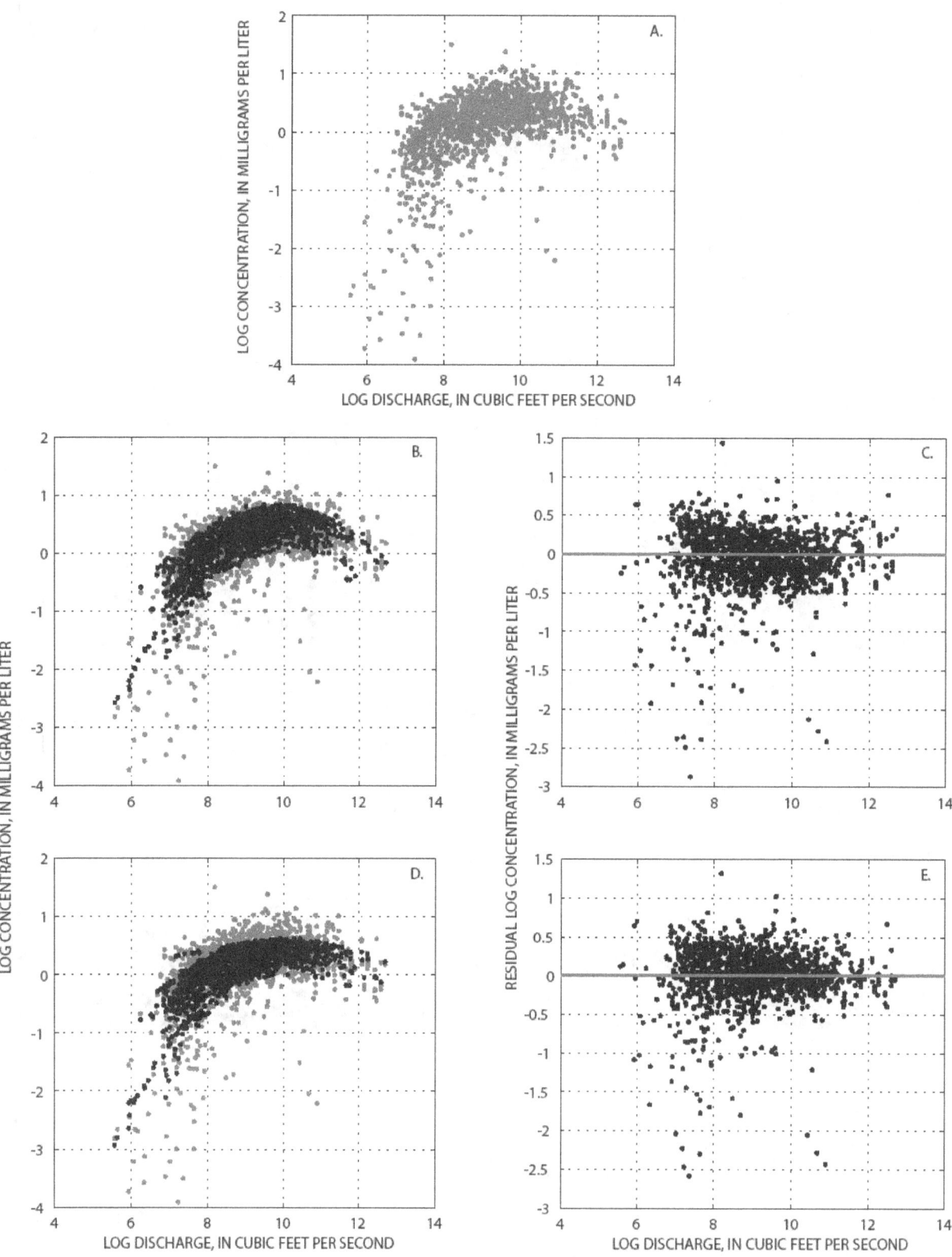

Figure 1–7.   Nitrate at Potomac River at Chain Bridge at Washington, D.C. (USGS Station ID 01646580), showing the *(A)* observed concentration (red dots) versus discharge relation, *(B)* observed (red dots) and ESTIMATOR-predicted (black dots) concentration versus discharge relation, *(C)* residual (observed minus predicted) plot for ESTIMATOR predictions, *(D)* observed (red dots) and WRTDS-predicted (black dots) concentration versus discharge relation, and *(E)* residual (observed minus predicted) plot for WRTDS predictions.

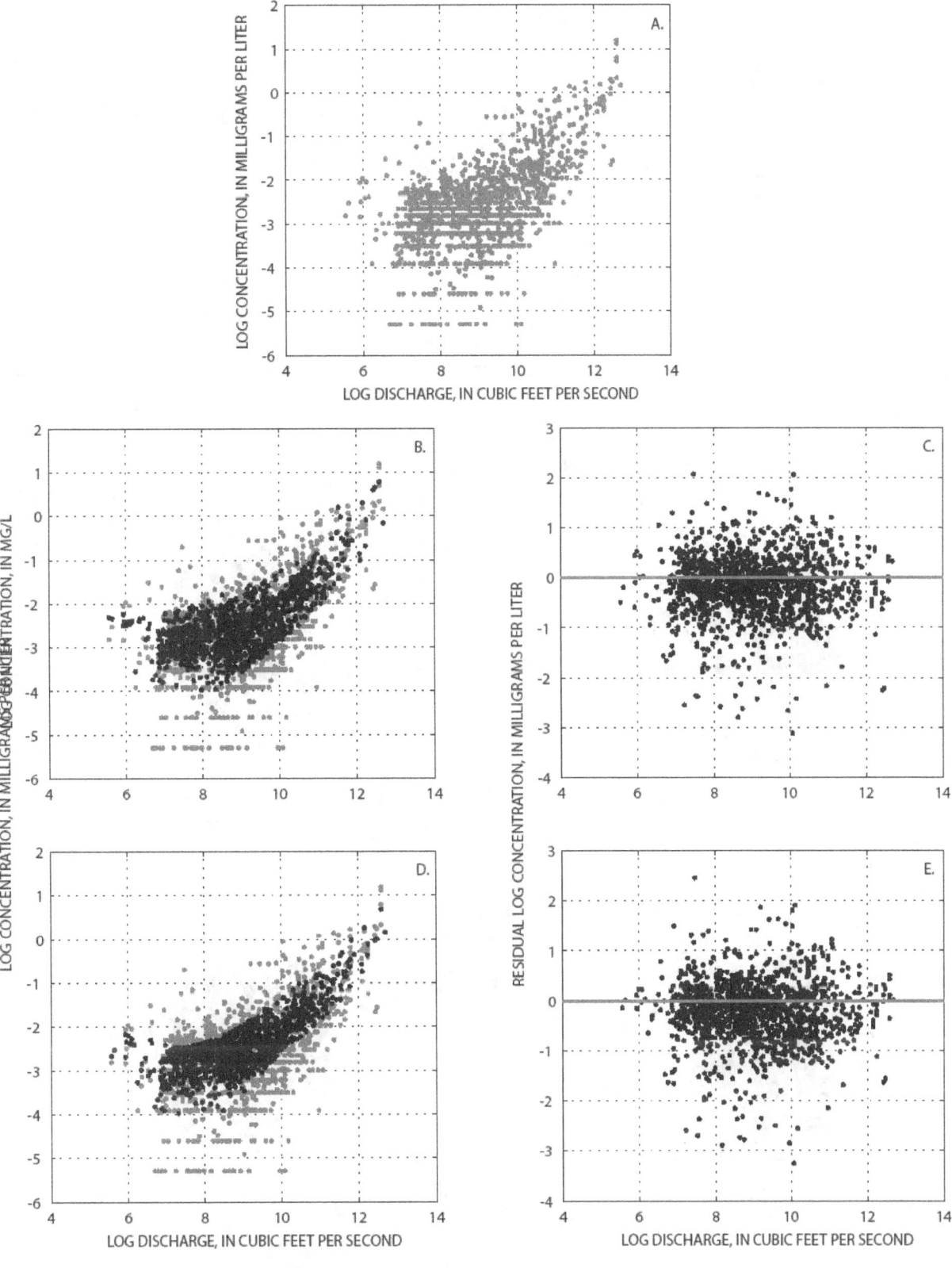

Figure 1–8. Total phosphorus at Potomac River at Chain Bridge at Washington, D.C. (USGS Station ID 01646580), showing the *(A)* observed concentration (red dots) versus discharge relation, *(B)* observed (red dots) and ESTIMATOR-predicted (black dots) concentration versus discharge relation, *(C)* residual (observed minus predicted) plot for ESTIMATOR predictions, *(D)* observed (red dots) and WRTDS-predicted (black dots) concentration versus discharge relation, and *(E)* residual (observed minus predicted) plot for WRTDS predictions.

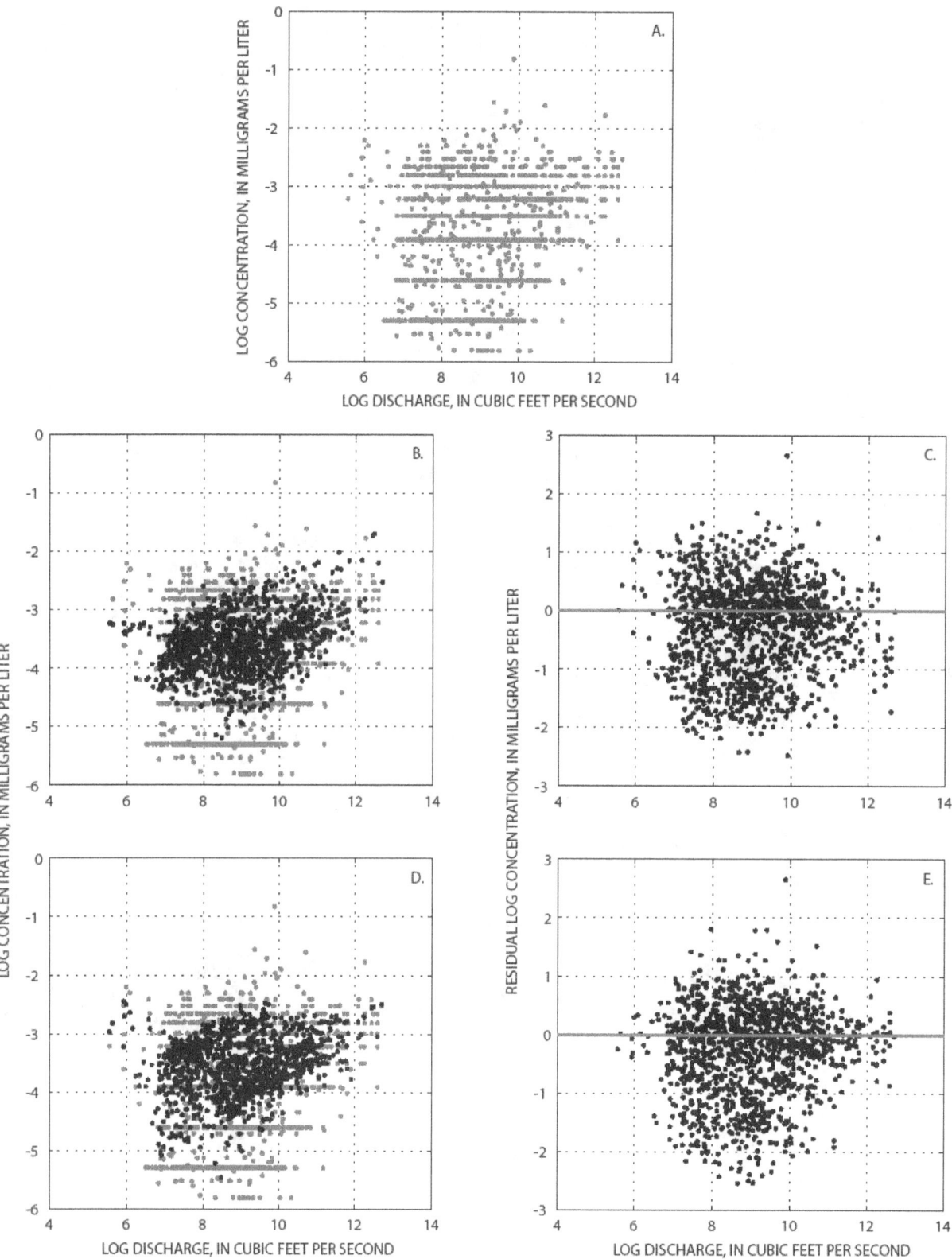

Figure 1– 9.   Orthophosphorus at Potomac River at Chain Bridge at Washington, D.C. (USGS Station ID 01646580), showing the *(A)* observed concentration (red dots) versus discharge relation, *(B)* observed (red dots) and ESTIMATOR-predicted (black dots) concentration versus discharge relation, *(C)* residual (observed minus predicted) plot for ESTIMATOR predictions, *(D)* observed (red dots) and WRTDS-predicted (black dots) concentration versus discharge relation, and *(E)* residual (observed minus predicted) plot for WRTDS predictions.

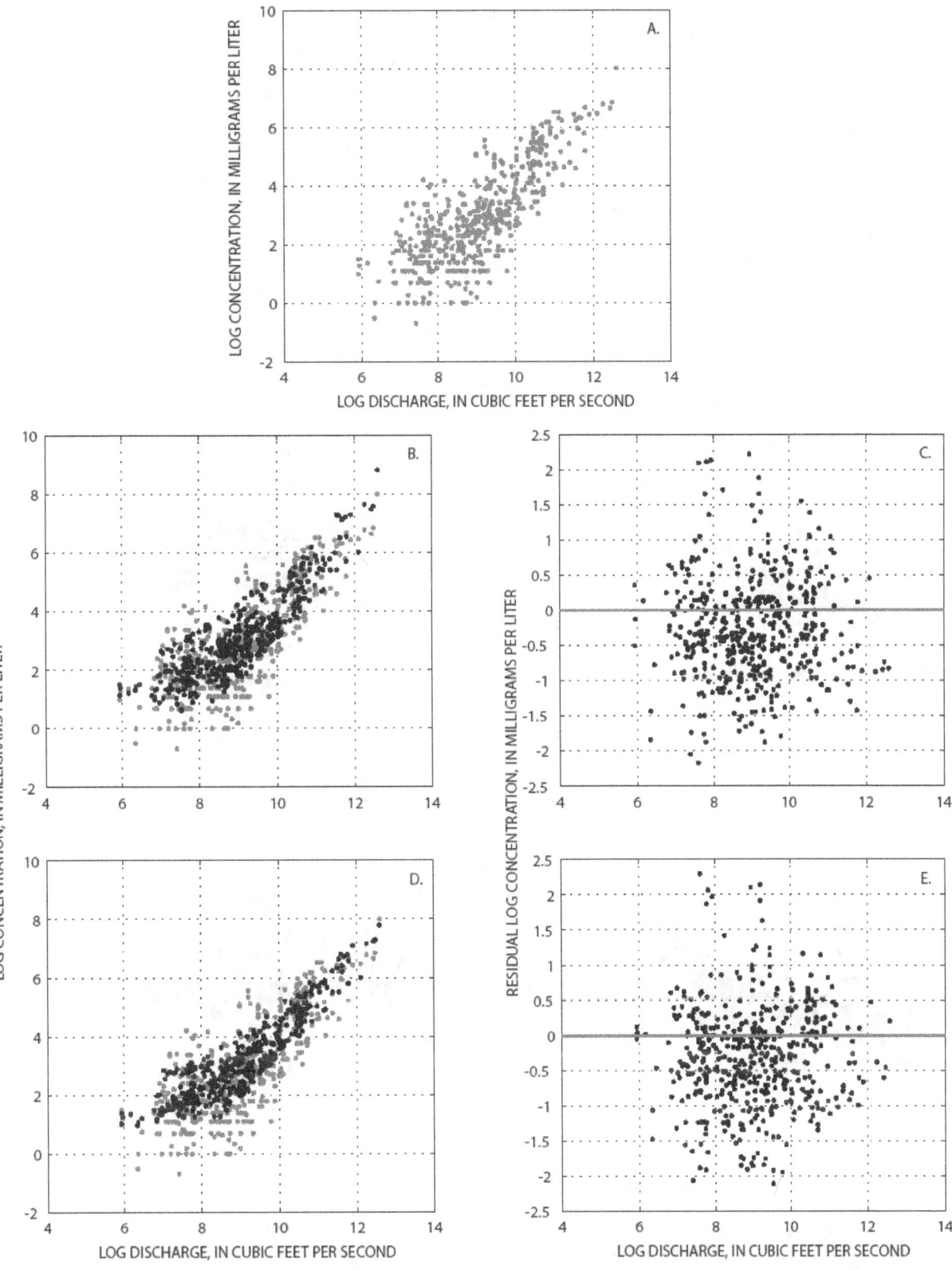

Figure 1–10. Suspended sediment at Potomac River at Chain Bridge at Washington, D.C. (USGS Station ID 01646580), showing the *(A)* observed concentration (red dots) versus discharge relation, *(B)* observed (red dots) and ESTIMATOR-predicted (black dots) concentration versus discharge relation, *(C)* residual (observed minus predicted) plot for ESTIMATOR predictions, *(D)* observed (red dots) and WRTDS-predicted (black dots) concentration versus discharge relation, and *(E)* residual (observed minus predicted) plot for WRTDS predictions.

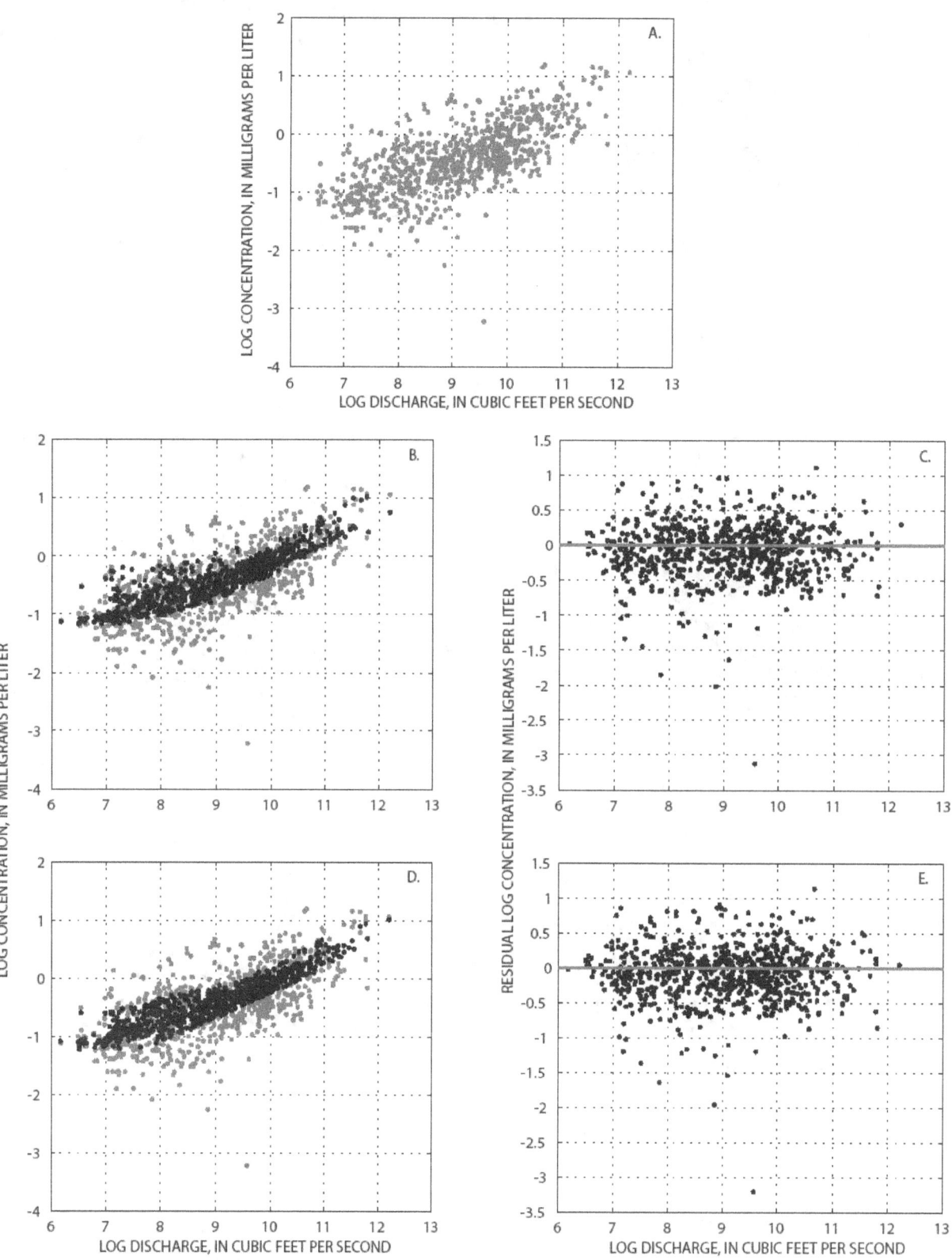

Figure 1–11.   Total nitrogen at James River at Cartersville, Virginia (USGS Station ID 02035000), showing the
(A) observed concentration (red dots) versus discharge relation, *(B)* observed (red dots) and ESTIMATOR-predicted
(black dots) concentration versus discharge relation, *(C)* residual (observed minus predicted) plot for ESTIMATOR
predictions, *(D)* observed (red dots) and WRTDS-predicted (black dots) concentration versus discharge relation, and
(E) residual (observed minus predicted) plot for WRTDS predictions.

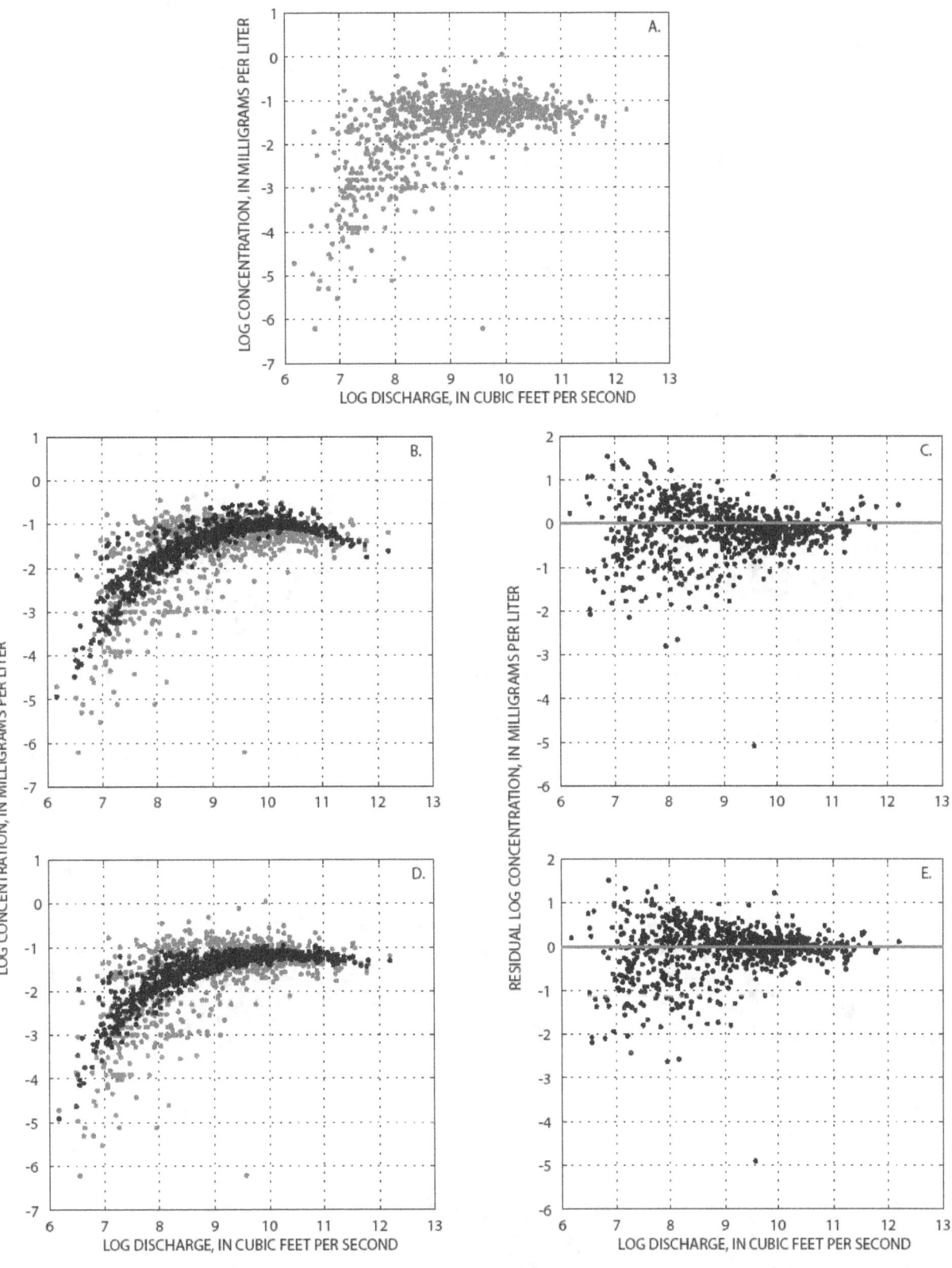

Figure 1–12.   Nitrate at James River at Cartersville, Virginia (USGS Station ID 02035000), showing the *(A)* observed concentration (red dots) versus discharge relation, *(B)* observed (red dots) and ESTIMATOR-predicted (black dots) concentration versus discharge relation, *(C)* residual (observed minus predicted) plot for ESTIMATOR predictions, *(D)* observed (red dots) and WRTDS-predicted (black dots) concentration versus discharge relation, and *(E)* residual (observed minus predicted) plot for WRTDS predictions.

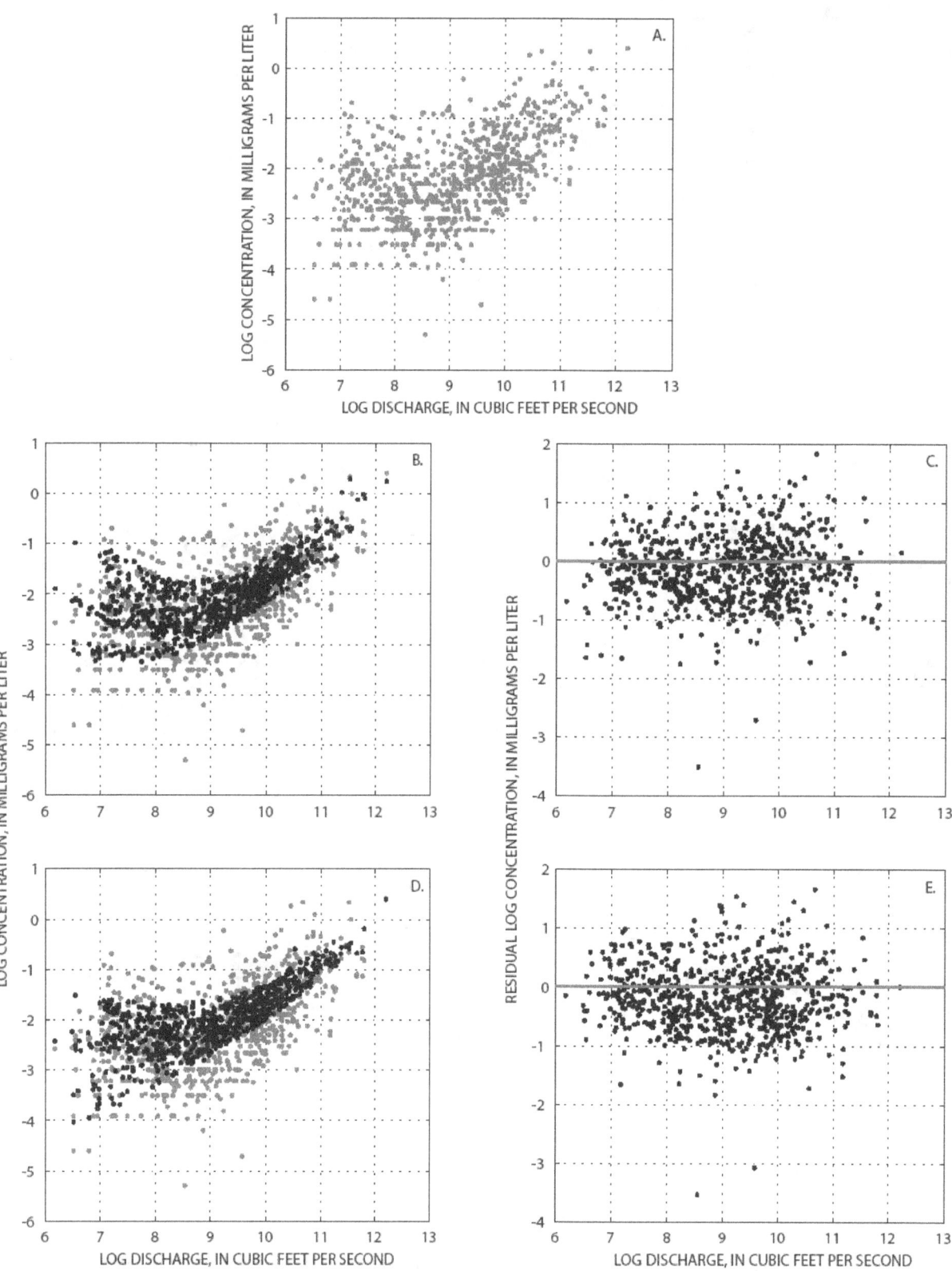

Figure 1–13.   Total phosphorus at James River at Cartersville, Virginia (USGS Station ID 02035000), showing the *(A)* observed concentration (red dots) versus discharge relation, *(B)* observed (red dots) and ESTIMATOR-predicted (black dots) concentration versus discharge relation, *(C)* residual (observed minus predicted) plot for ESTIMATOR predictions, *(D)* observed (red dots) and WRTDS-predicted (black dots) concentration versus discharge relation, and *(E)* residual (observed minus predicted) plot for WRTDS predictions.

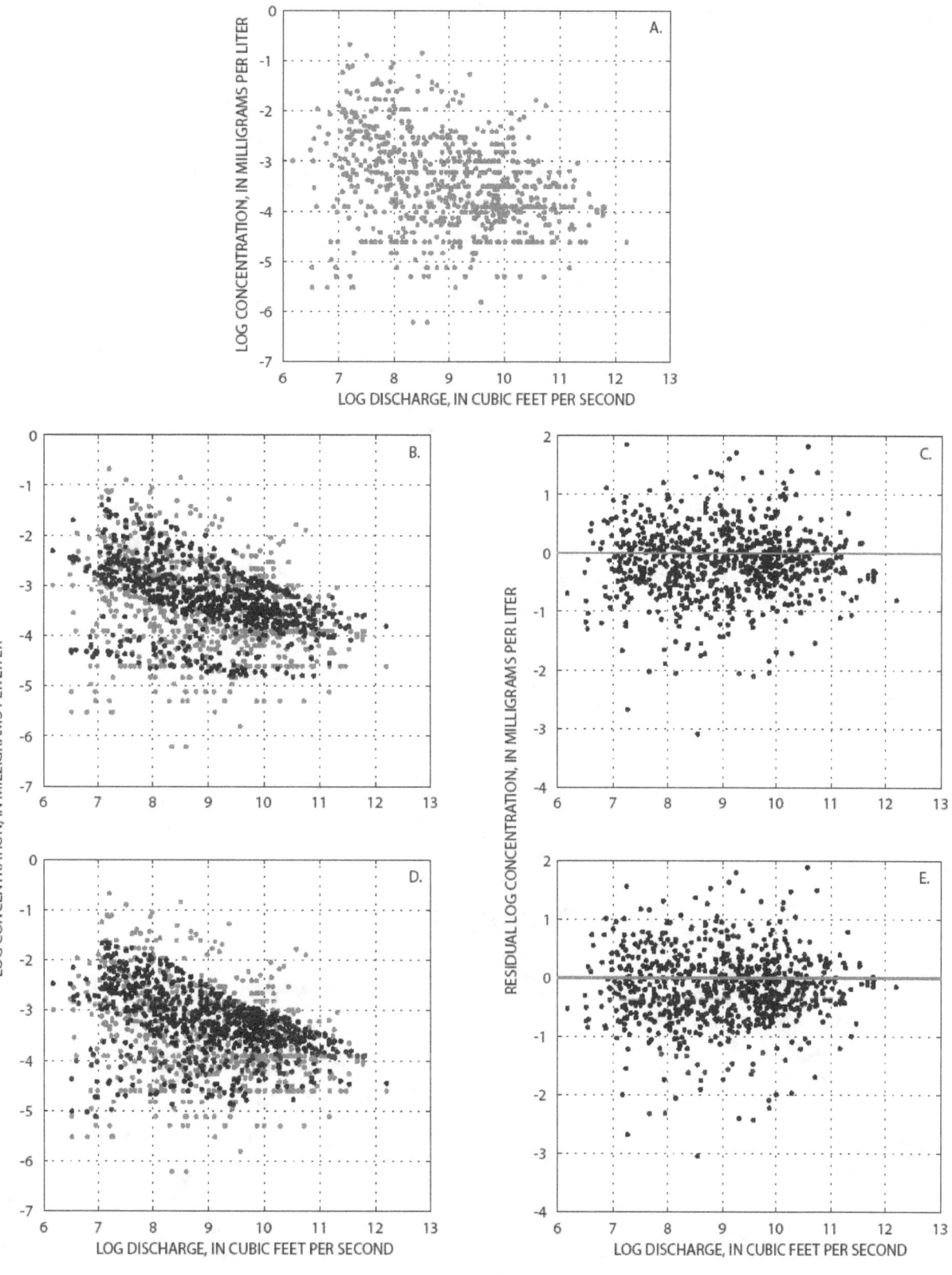

Figure 1–14. Orthophosphorus at James River at Cartersville, Virginia (USGS Station ID 02035000), showing the *(A)* observed concentration (red dots) versus discharge relation, *(B)* observed (red dots) and ESTIMATOR-predicted (black dots) concentration versus discharge relation, *(C)* residual (observed minus predicted) plot for ESTIMATOR predictions, *(D)* observed (red dots) and WRTDS-predicted (black dots) concentration versus discharge relation, and *(E)* residual (observed minus predicted) plot for WRTDS predictions.

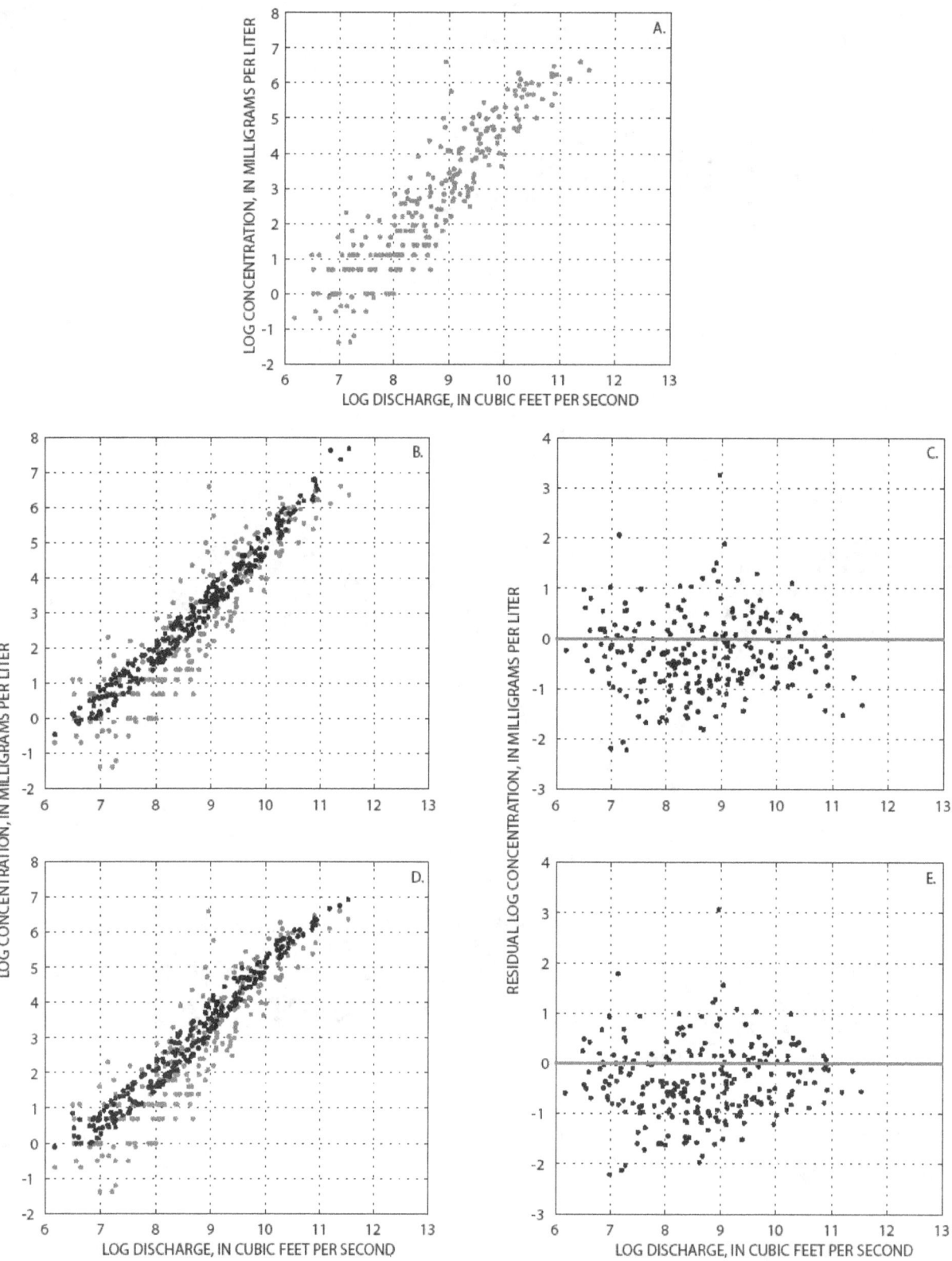

Figure 1–15.   Suspended sediment at James River at Cartersville, Virginia (USGS Station ID 02035000), showing the *(A)* observed concentration (red dots) versus discharge relation, *(B)* observed (red dots) and ESTIMATOR-predicted (black dots) concentration versus discharge relation, *(C)* residual (observed minus predicted) plot for ESTIMATOR predictions, *(D)* observed (red dots) and WRTDS-predicted (black dots) concentration versus discharge relation, and *(E)* residual (observed minus predicted) plot for WRTDS predictions.

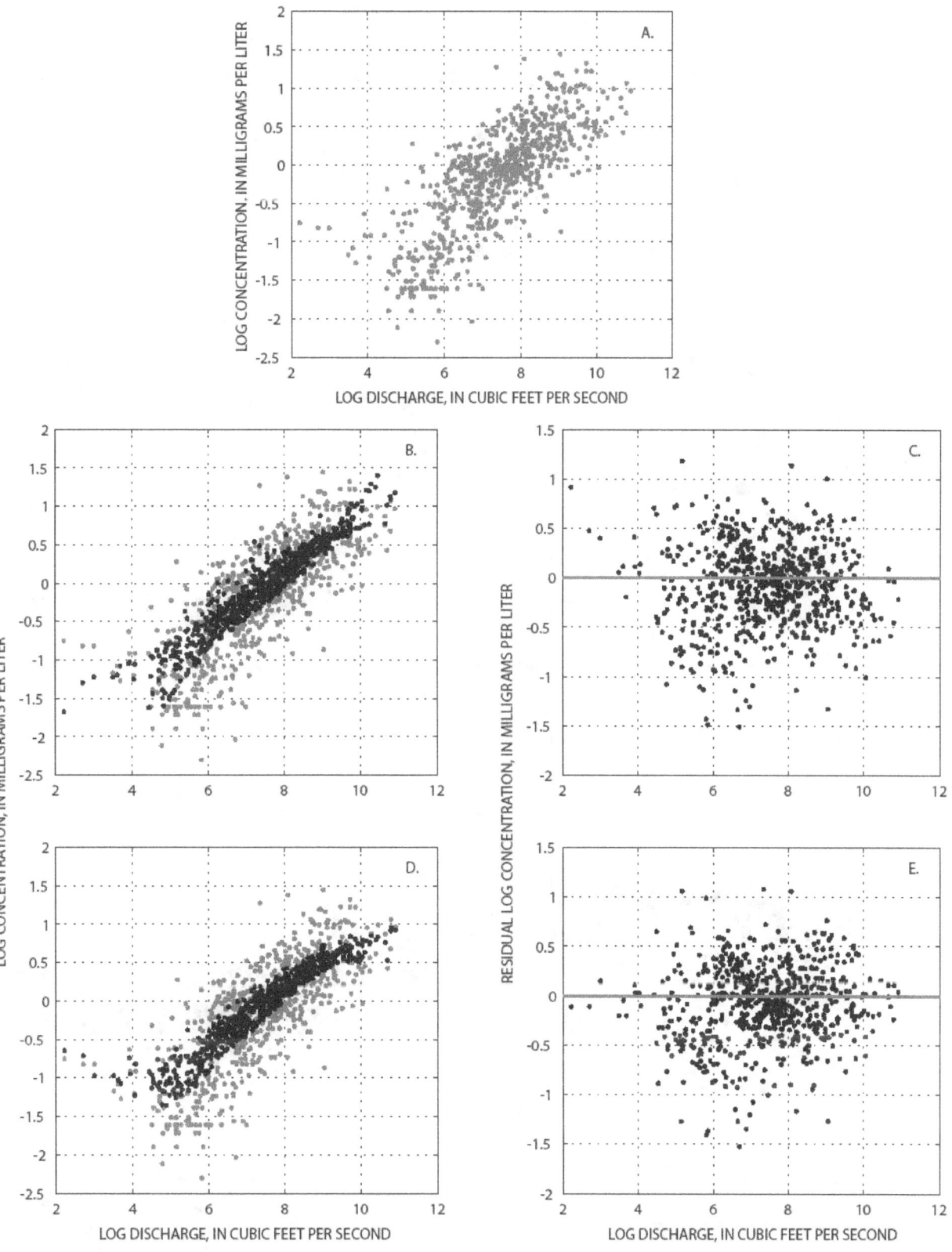

Figure 1–16. Total nitrogen at Rappahannock River near Fredericksburg, Virginia (USGS Station ID 01668000), showing the *(A)* observed concentration (red dots) versus discharge relation, *(B)* observed (red dots) and ESTIMATOR-predicted (black dots) concentration versus discharge relation, *(C)* residual (observed minus predicted) plot for ESTIMATOR predictions, *(D)* observed (red dots) and WRTDS-predicted (black dots) concentration versus discharge relation, and *(E)* residual (observed minus predicted) plot for WRTDS predictions.

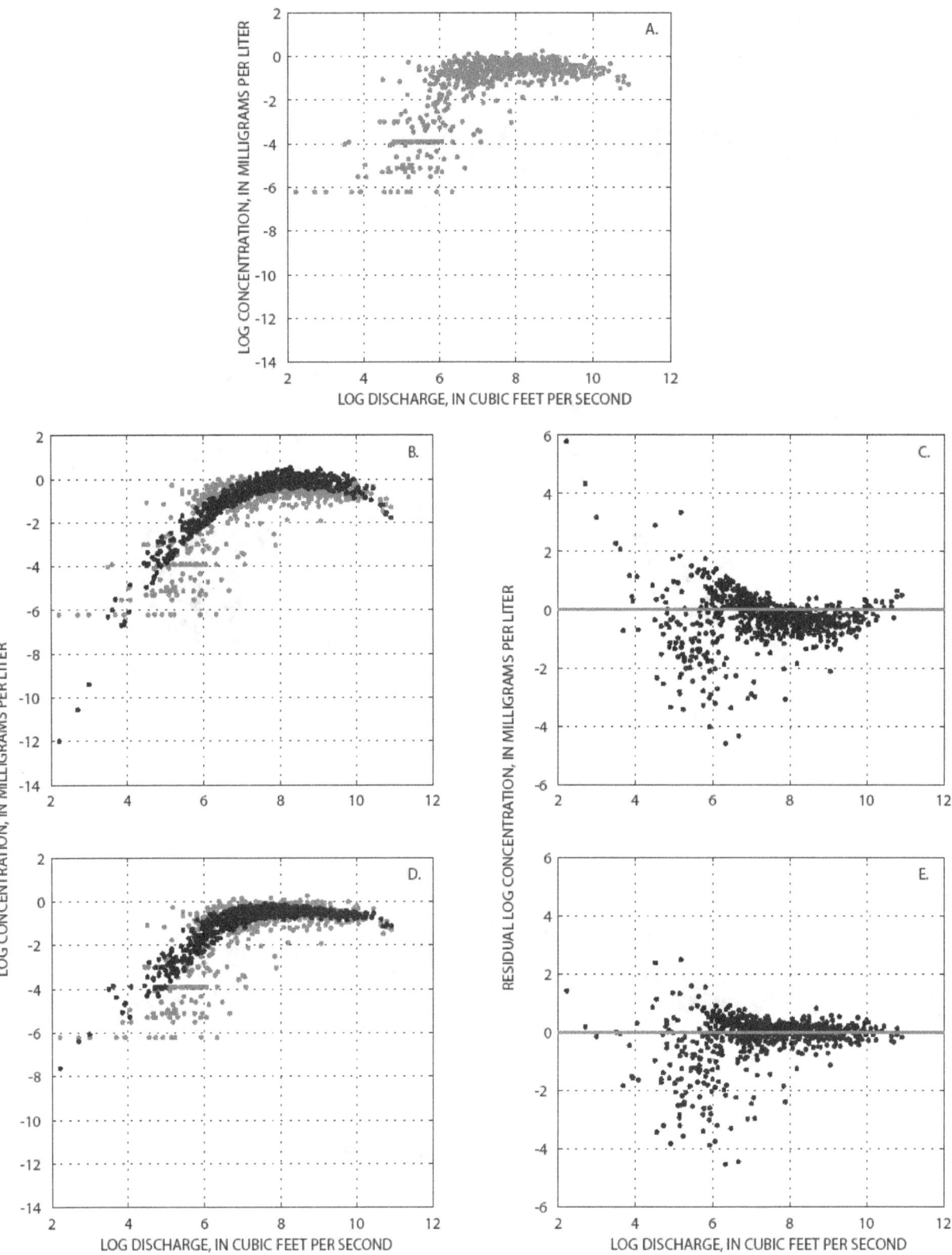

Figure 1–17. Nitrate at Rappahannock River near Fredericksburg, Virginia (USGS Station ID 01668000), showing the *(A)* observed concentration (red dots) versus discharge relation, *(B)* observed (red dots) and ESTIMATOR-predicted (black dots) concentration versus discharge relation, *(C)* residual (observed minus predicted) plot for ESTIMATOR predictions, *(D)* observed (red dots) and WRTDS-predicted (black dots) concentration versus discharge relation, and *(E)* residual (observed minus predicted) plot for WRTDS predictions.

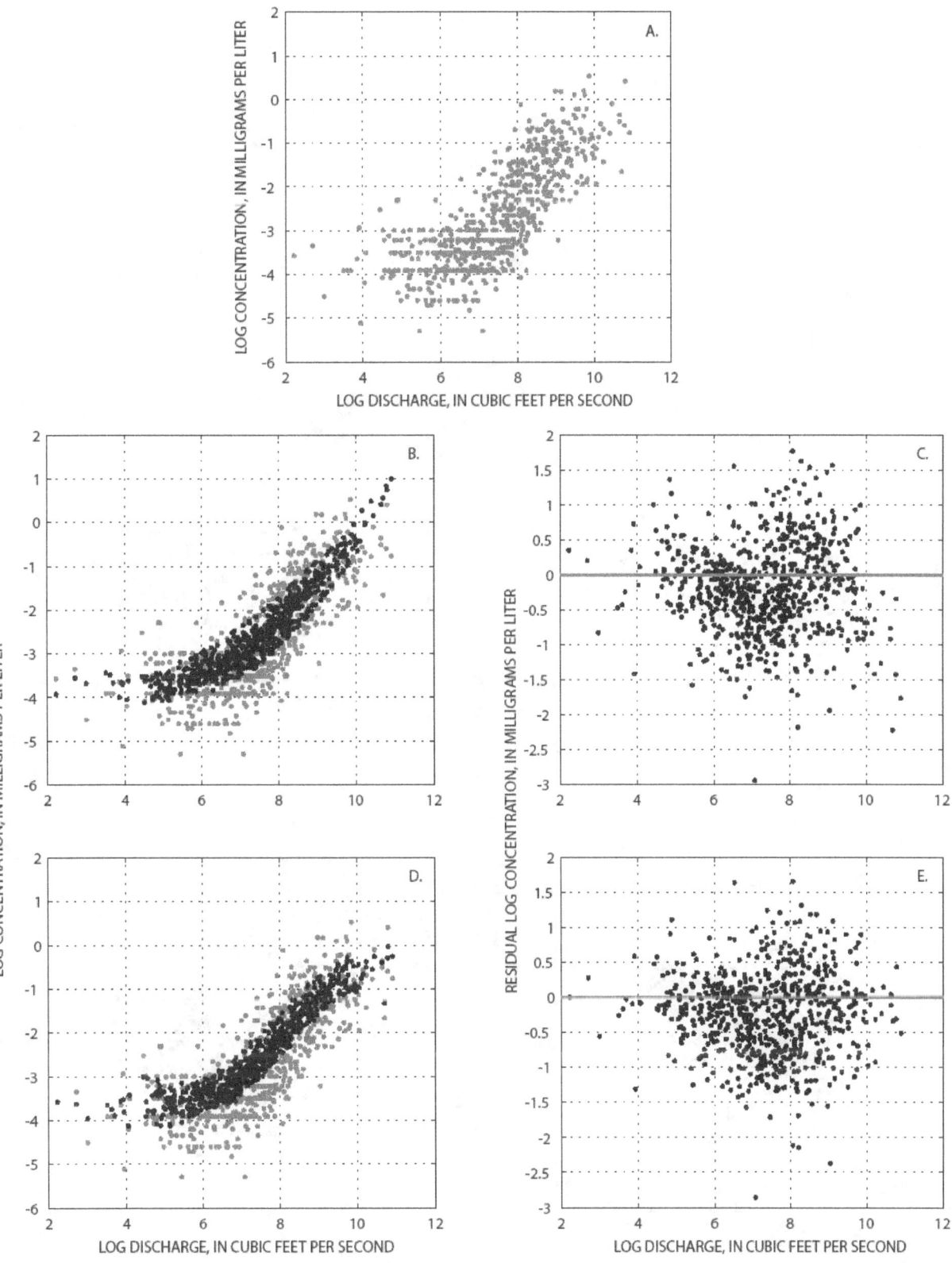

Figure 1–18. Total phosphorus at Rappahannock River near Fredericksburg, Virginia (USGS Station ID 01668000), showing the *(A)* observed concentration (red dots) versus discharge relation, *(B)* observed (red dots) and ESTIMATOR-predicted (black dots) concentration versus discharge relation, *(C)* residual (observed minus predicted) plot for ESTIMATOR predictions, *(D)* observed (red dots) and WRTDS-predicted (black dots) concentration versus discharge relation, and *(E)* residual (observed minus predicted) plot for WRTDS predictions.

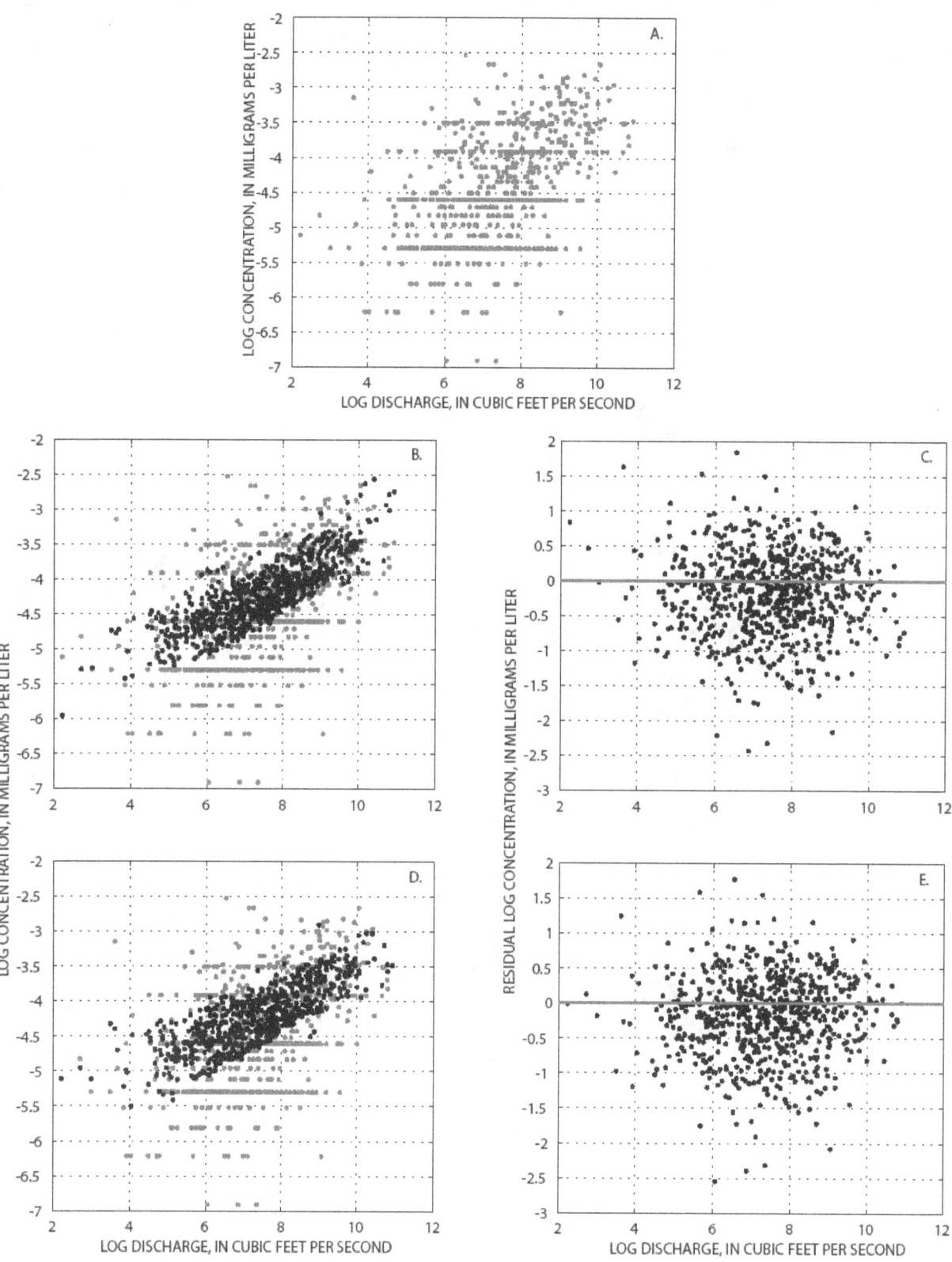

Figure 1–19.   Orthophosphorus at Rappahannock River near Fredericksburg, Virginia (USGS Station ID 01668000), showing the *(A)* observed concentration (red dots) versus discharge relation, *(B)* observed (red dots) and ESTIMATOR-predicted (black dots) concentration versus discharge relation, *(C)* residual (observed minus predicted) plot for ESTIMATOR predictions, *(D)* observed (red dots) and WRTDS-predicted (black dots) concentration versus discharge relation, and *(E)* residual (observed minus predicted) plot for WRTDS predictions.

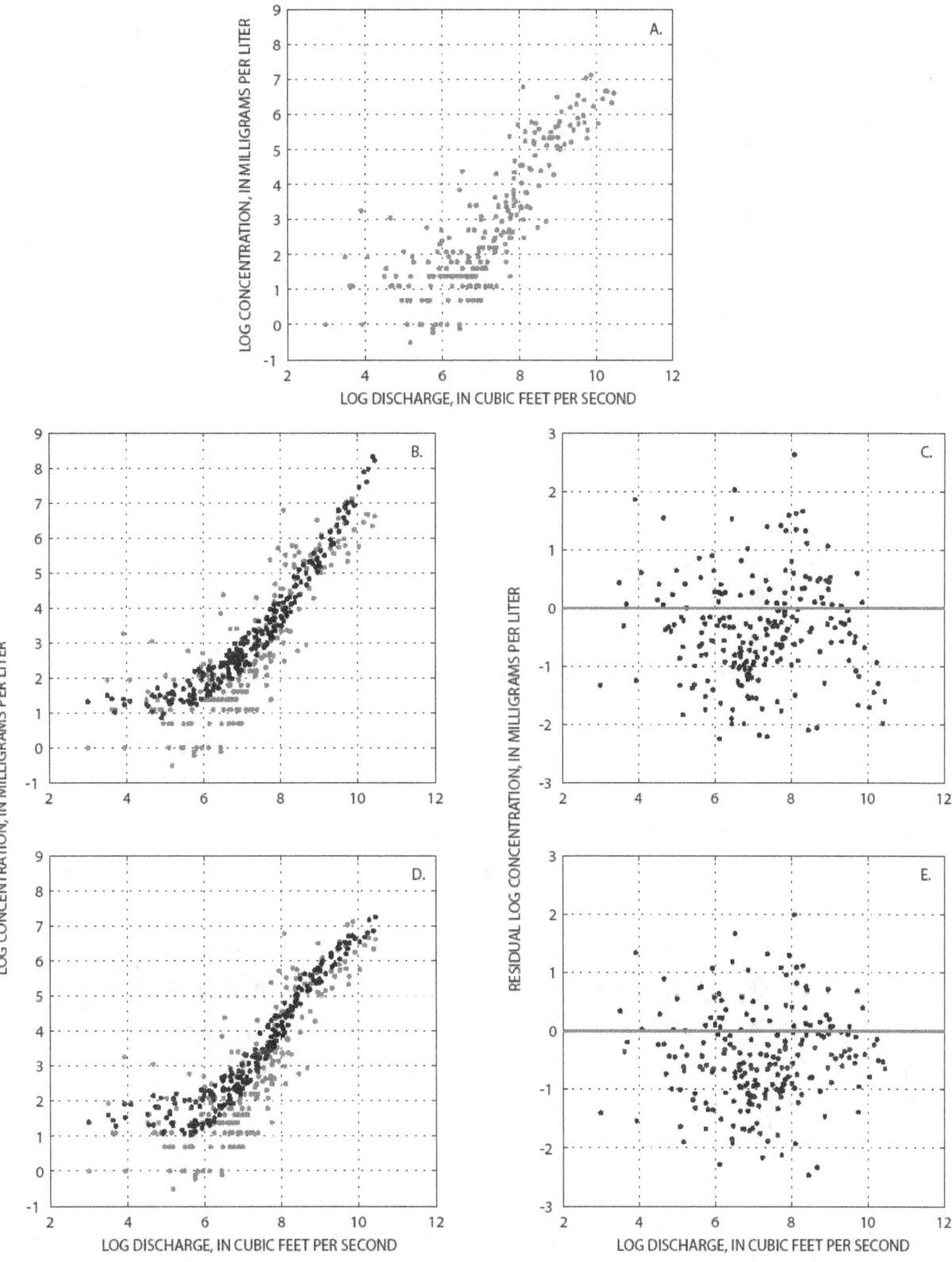

Figure 1–20.   Suspended sediment at Rappahannock River near Fredericksburg, Virginia (USGS Station ID 01668000), showing the *(A)* observed concentration (red dots) versus discharge relation, *(B)* observed (red dots) and ESTIMATOR-predicted (black dots) concentration versus discharge relation, *(C)* residual (observed minus predicted) plot for ESTIMATOR predictions, *(D)* observed (red dots) and WRTDS-predicted (black dots) concentration versus discharge relation, and *(E)* residual (observed minus predicted) plot for WRTDS predictions.

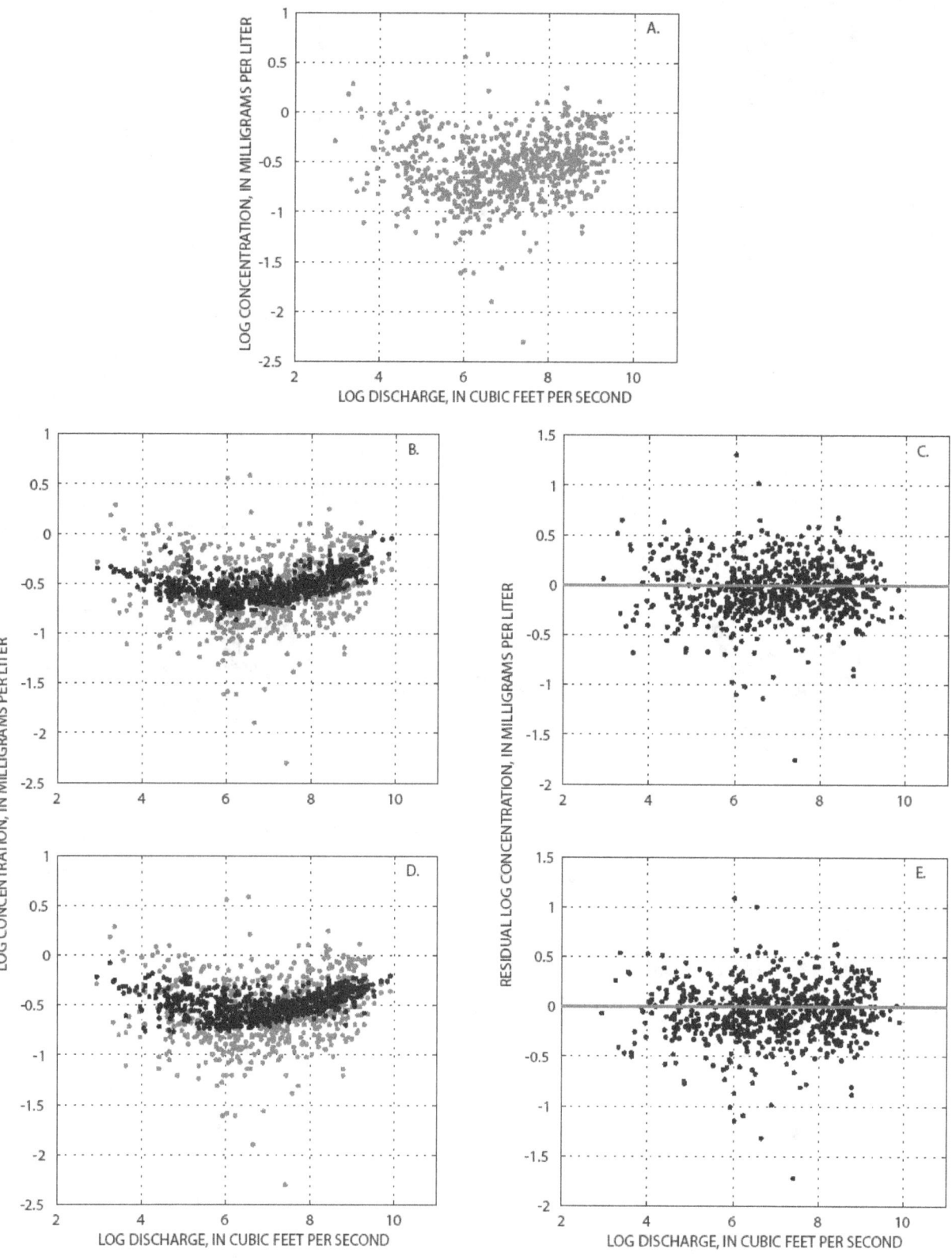

Figure 1–21.   Total nitrogen at Appomattox River near Matoaca, Virginia (USGS Station ID 02041650), showing the *(A)* observed concentration (red dots) versus discharge relation, *(B)* observed (red dots) and ESTIMATOR-predicted (black dots) concentration versus discharge relation, *(C)* residual (observed minus predicted) plot for ESTIMATOR predictions, *(D)* observed (red dots) and WRTDS-predicted (black dots) concentration versus discharge relation, and *(E)* residual (observed minus predicted) plot for WRTDS predictions.

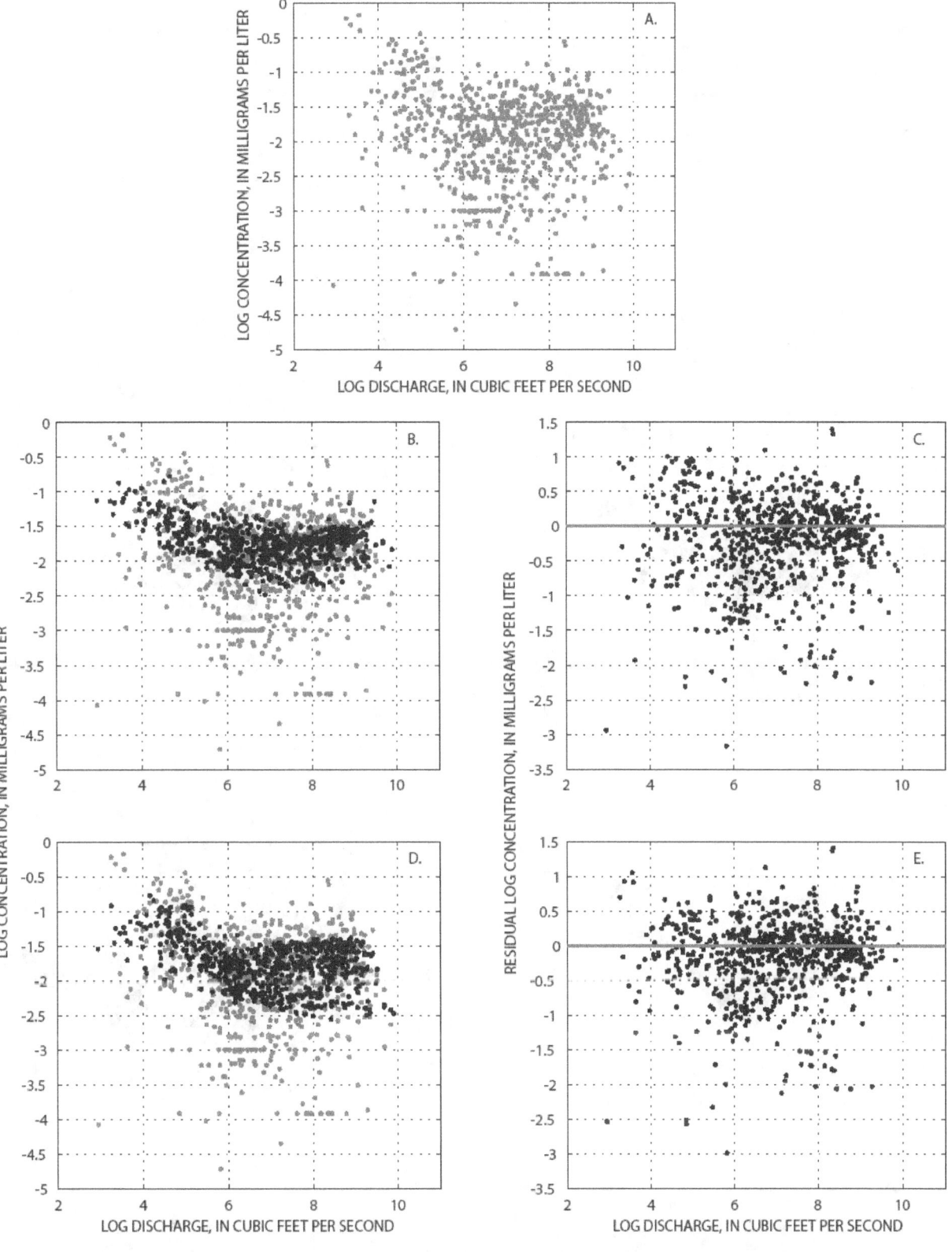

Figure 1–22.   Nitrate at Appomattox River near Matoaca, Virginia (USGS Station ID 02041650), showing the *(A)* observed concentration (red dots) versus discharge relation, *(B)* observed (red dots) and ESTIMATOR-predicted (black dots) concentration versus discharge relation, *(C)* residual (observed minus predicted) plot for ESTIMATOR predictions, *(D)* observed (red dots) and WRTDS-predicted (black dots) concentration versus discharge relation, and *(E)* residual (observed minus predicted) plot for WRTDS predictions.

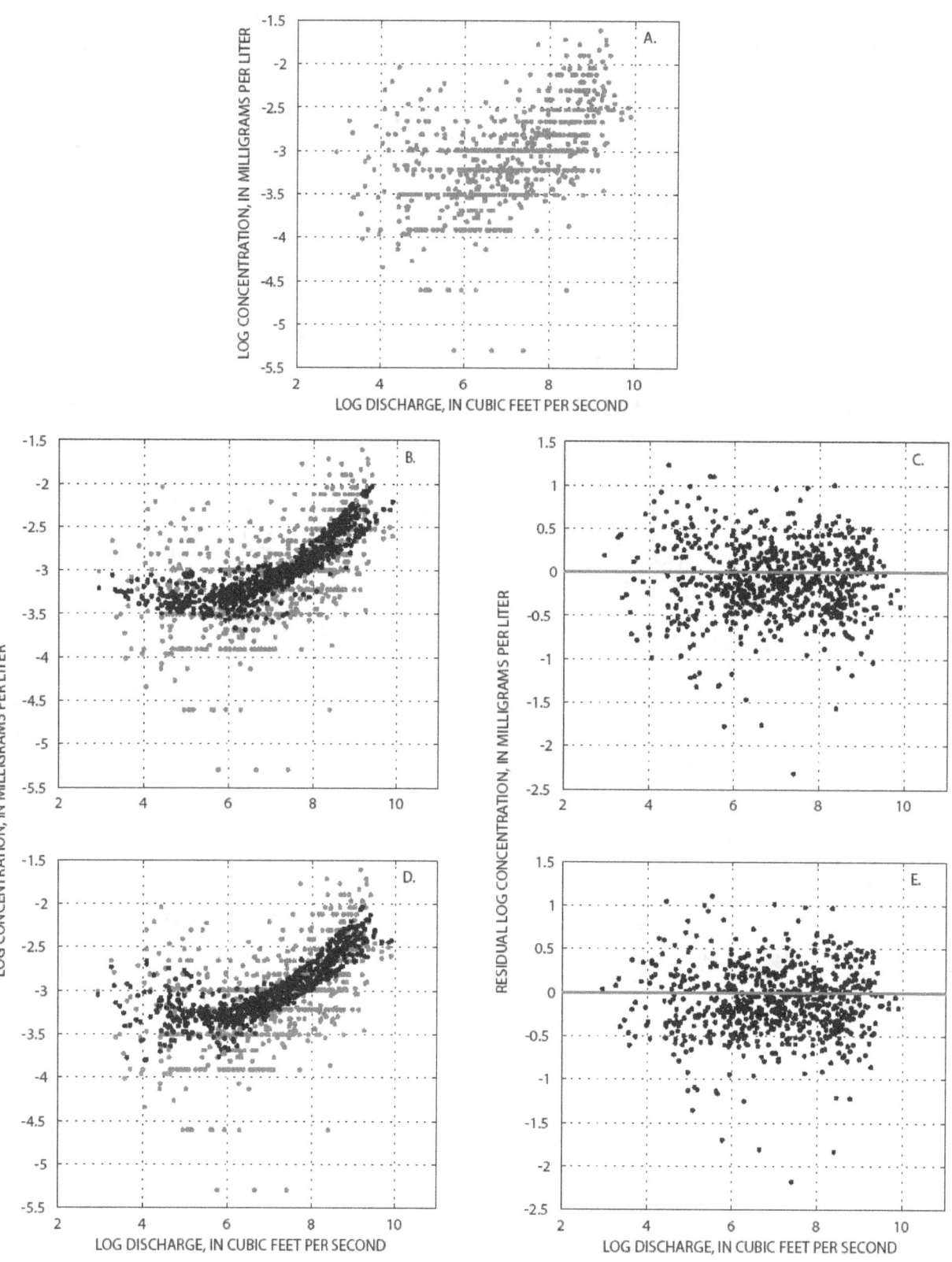

Figure 1–23. Total phosphorus at Appomattox River near Matoaca, Virginia (USGS Station ID 02041650), showing the *(A)* observed concentration (red dots) versus discharge relation, *(B)* observed (red dots) and ESTIMATOR-predicted (black dots) concentration versus discharge relation, *(C)* residual (observed minus predicted) plot for ESTIMATOR predictions, *(D)* observed (red dots) and WRTDS-predicted (black dots) concentration versus discharge relation, and *(E)* residual (observed minus predicted) plot for WRTDS predictions.

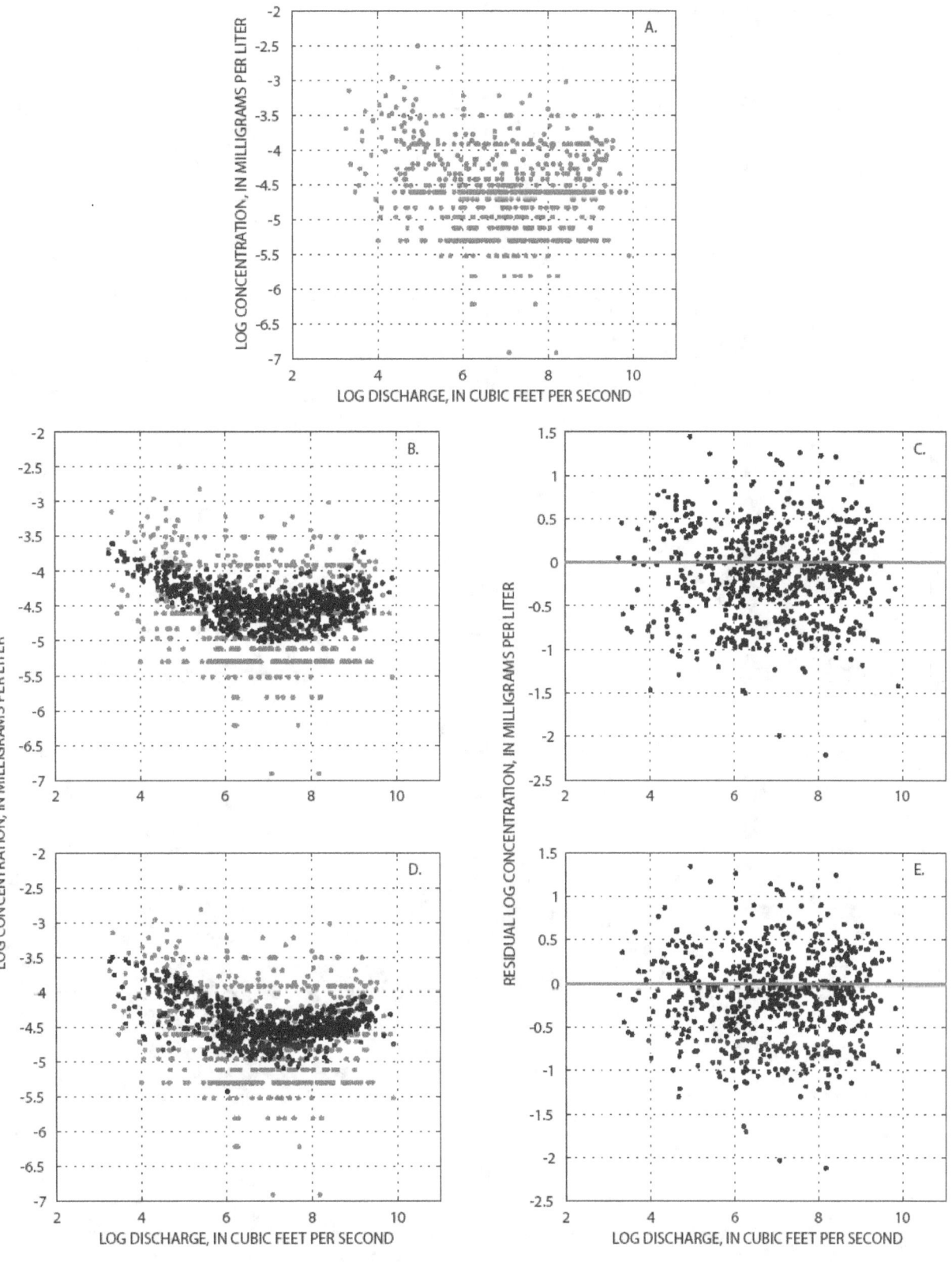

Figure 1–24. Orthophosphorus at Appomattox River near Matoaca, Virginia (USGS Station ID 02041650), showing the *(A)* observed concentration (red dots) versus discharge relation, *(B)* observed (red dots) and ESTIMATOR-predicted (black dots) concentration versus discharge relation, *(C)* residual (observed minus predicted) plot for ESTIMATOR predictions, *(D)* observed (red dots) and WRTDS-predicted (black dots) concentration versus discharge relation, and *(E)* residual (observed minus predicted) plot for WRTDS predictions.

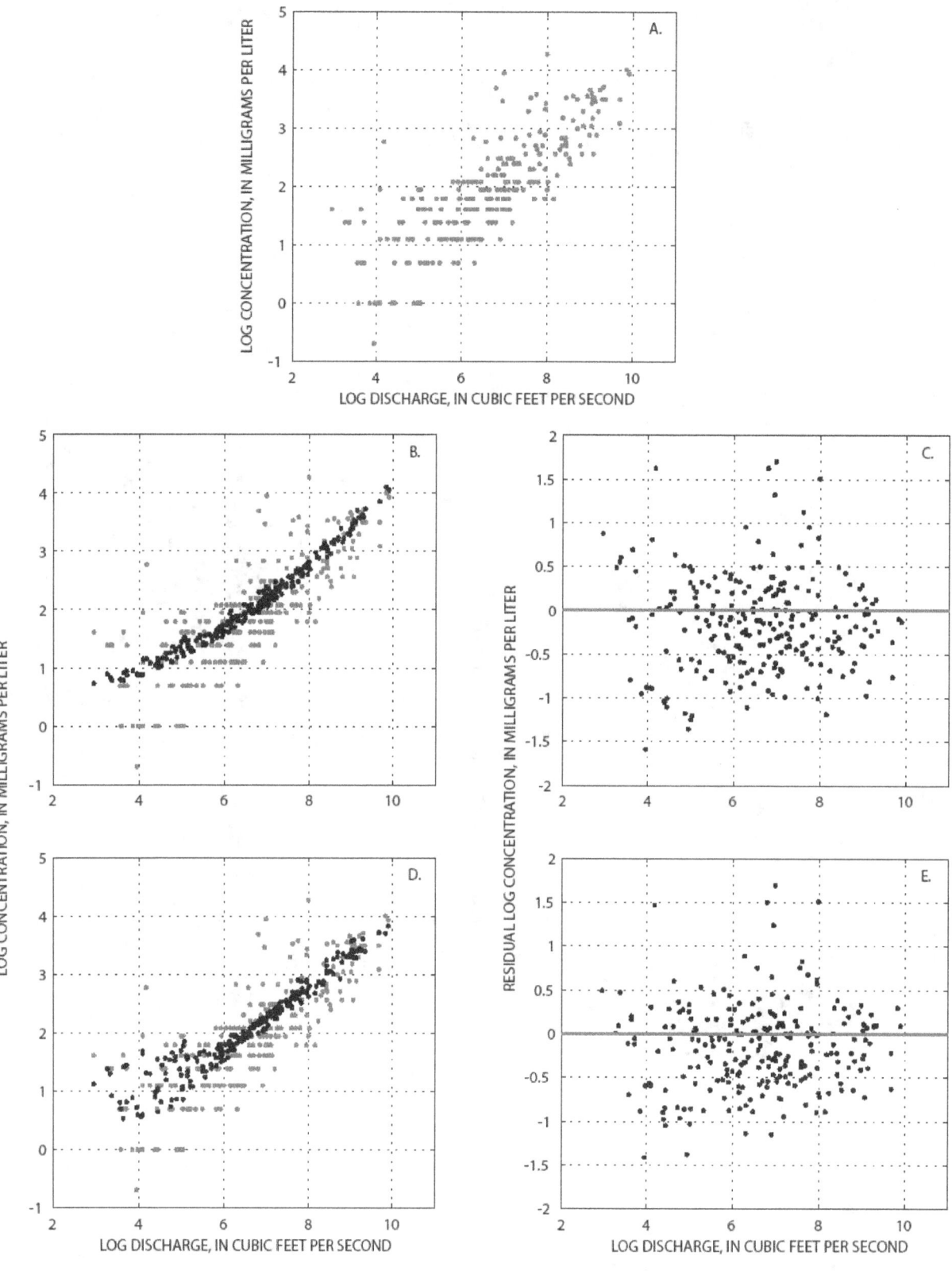

Figure 1–25.   Suspended sediment at Appomattox River near Matoaca, Virginia (USGS Station ID 02041650), showing the *(A)* observed concentration (red dots) versus discharge relation, *(B)* observed (red dots) and ESTIMATOR-predicted (black dots) concentration versus discharge relation, *(C)* residual (observed minus predicted) plot for ESTIMATOR predictions, *(D)* observed (red dots) and WRTDS-predicted (black dots) concentration versus discharge relation, and *(E)* residual (observed minus predicted) plot for WRTDS predictions.

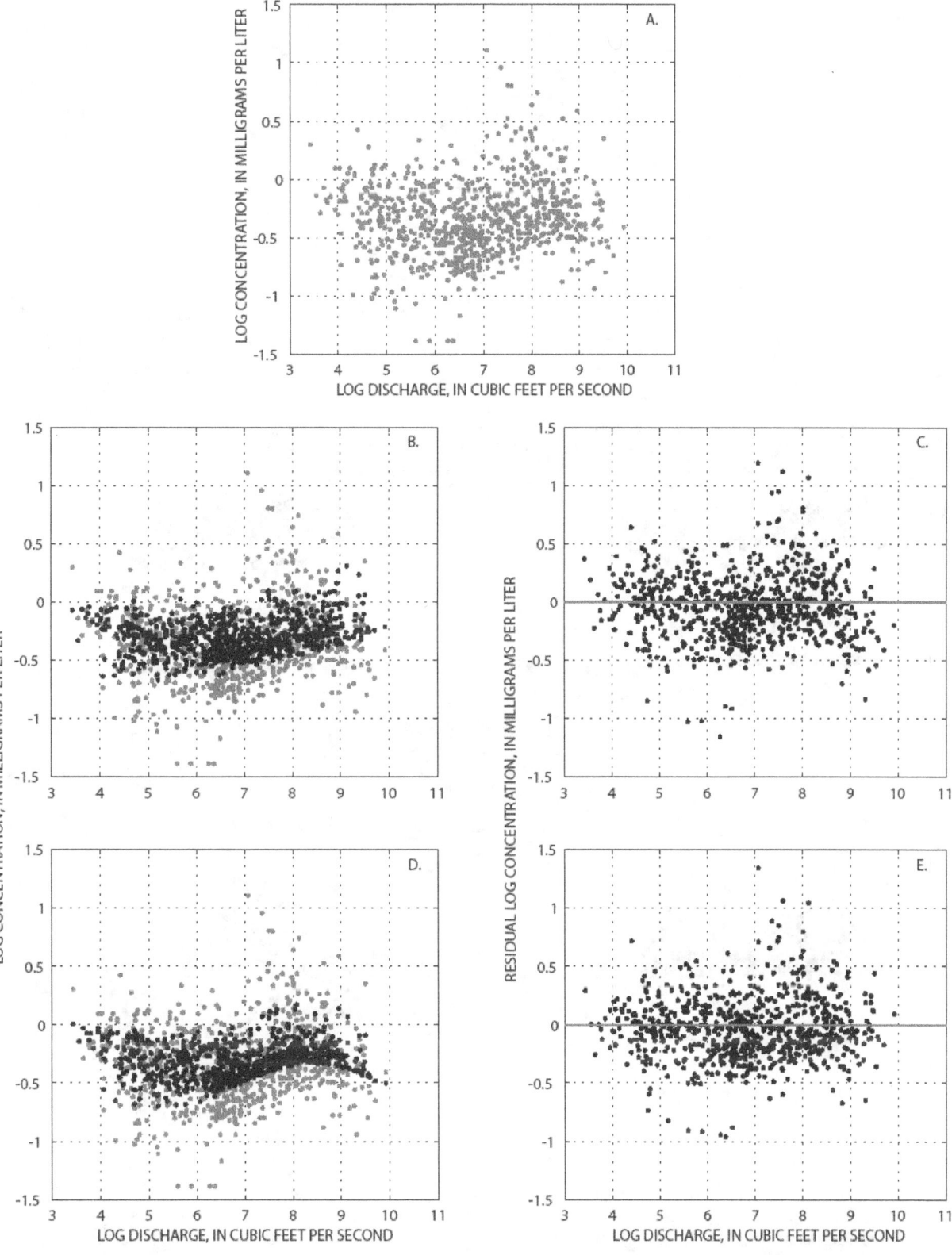

Figure 1–26. Total nitrogen at Pamunkey River near Hanover, Virginia (USGS Station ID 01673000), showing the *(A)* observed concentration (red dots) versus discharge relation, *(B)* observed (red dots) and ESTIMATOR-predicted (black dots) concentration versus discharge relation, *(C)* residual (observed minus predicted) plot for ESTIMATOR predictions, *(D)* observed (red dots) and WRTDS-predicted (black dots) concentration versus discharge relation, and *(E)* residual (observed minus predicted) plot for WRTDS predictions.

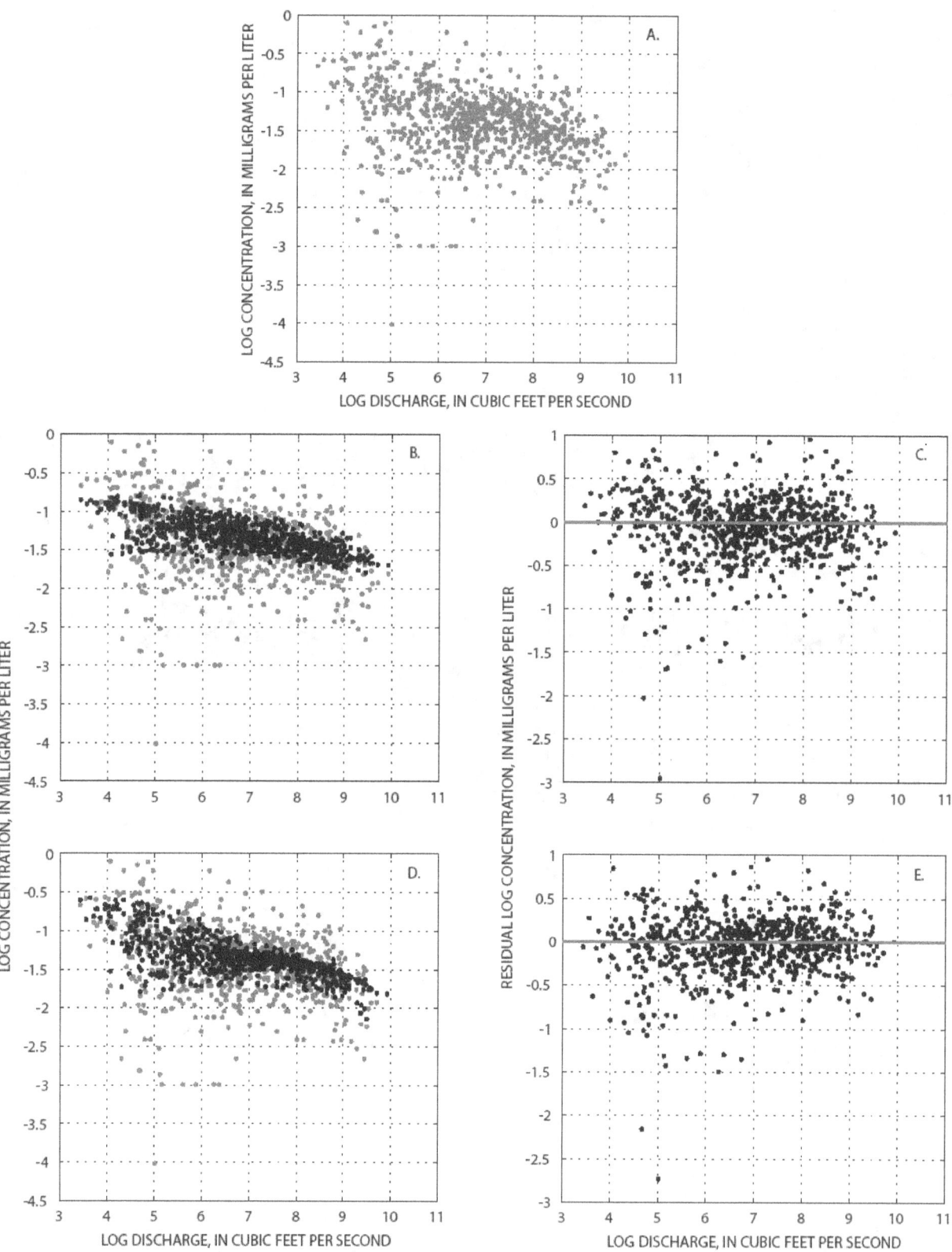

Figure 1–27.   Nitrate at Pamunkey River near Hanover, Virginia (USGS Station ID 01673000), showing the *(A)* observed concentration (red dots) versus discharge relation, *(B)* observed (red dots) and ESTIMATOR-predicted (black dots) concentration versus discharge relation, *(C)* residual (observed minus predicted) plot for ESTIMATOR predictions, *(D)* observed (red dots) and WRTDS-predicted (black dots) concentration versus discharge relation, and *(E)* residual (observed minus predicted) plot for WRTDS predictions.

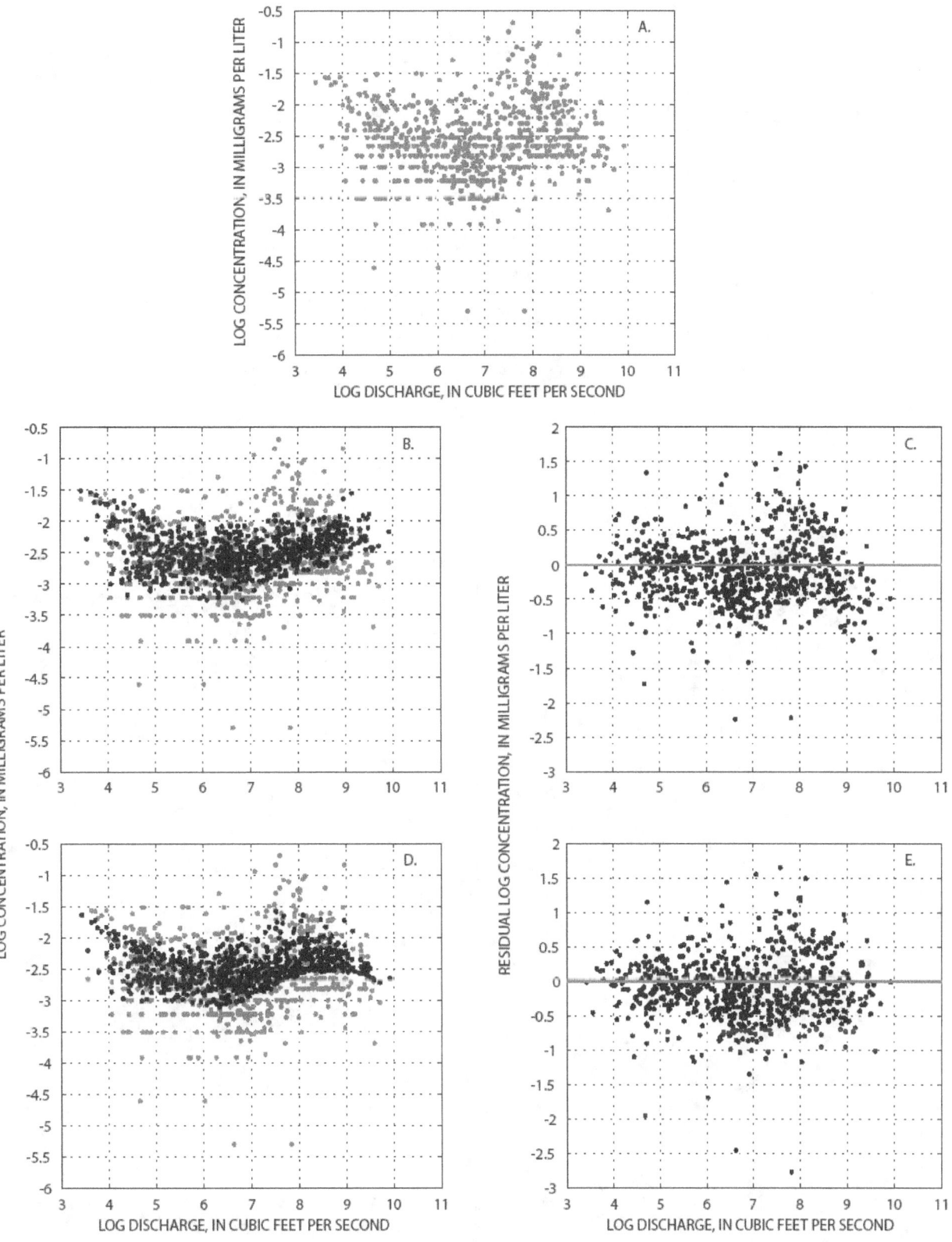

Figure 1–28. Total phosphorus at Pamunkey River near Hanover, Virginia (USGS Station ID 01673000), showing the *(A)* observed concentration (red dots) versus discharge relation, *(B)* observed (red dots) and ESTIMATOR-predicted (black dots) concentration versus discharge relation, *(C)* residual (observed minus predicted) plot for ESTIMATOR predictions, *(D)* observed (red dots) and WRTDS-predicted (black dots) concentration versus discharge relation, and *(E)* residual (observed minus predicted) plot for WRTDS predictions.

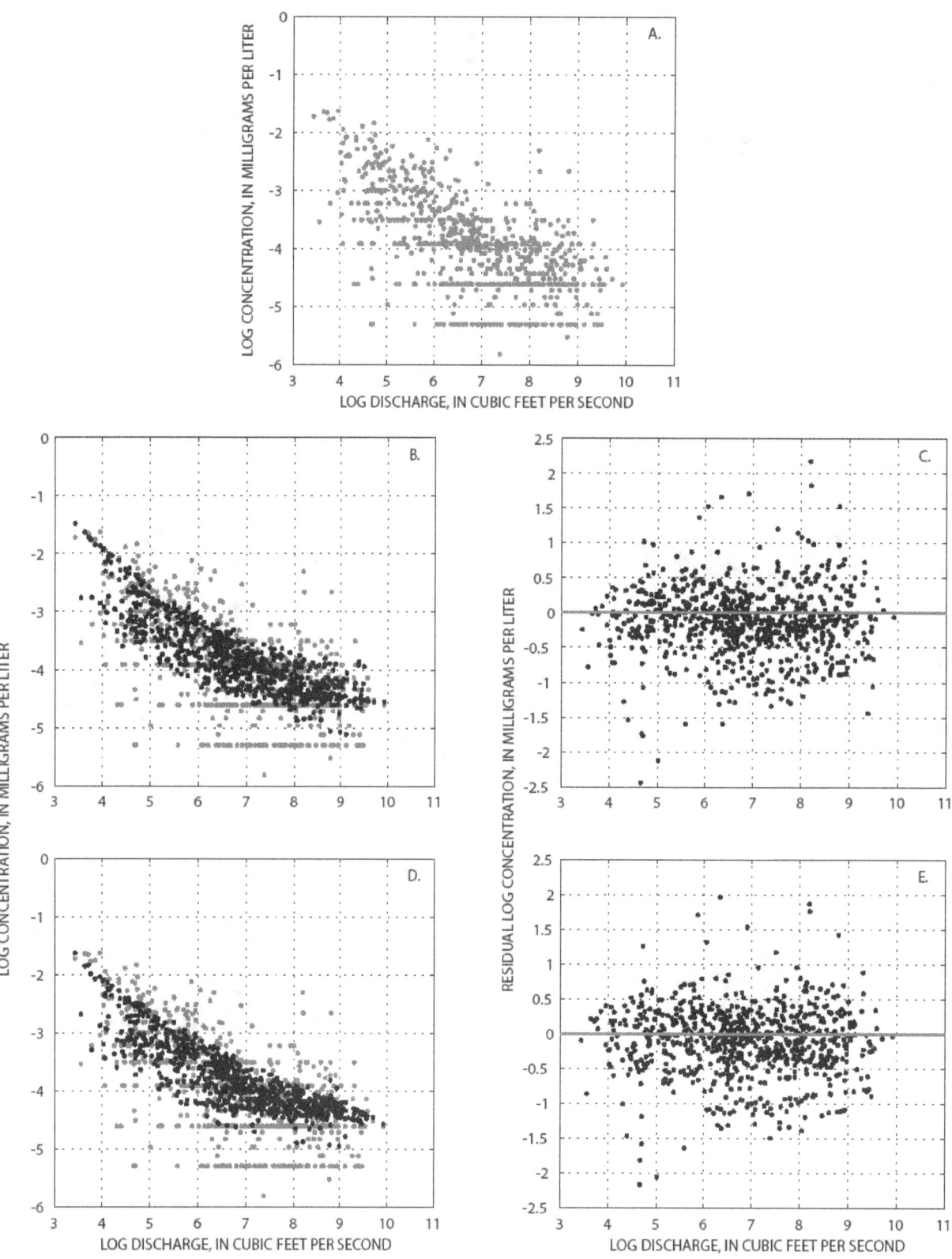

Figure 1–29.   Orthophosphorus at Pamunkey River near Hanover, Virginia (USGS Station ID 01673000), showing the *(A)* observed concentration (red dots) versus discharge relation, *(B)* observed (red dots) and ESTIMATOR-predicted (black dots) concentration versus discharge relation, *(C)* residual (observed minus predicted) plot for ESTIMATOR predictions, *(D)* observed (red dots) and WRTDS-predicted (black dots) concentration versus discharge relation, and *(E)* residual (observed minus predicted) plot for WRTDS predictions.

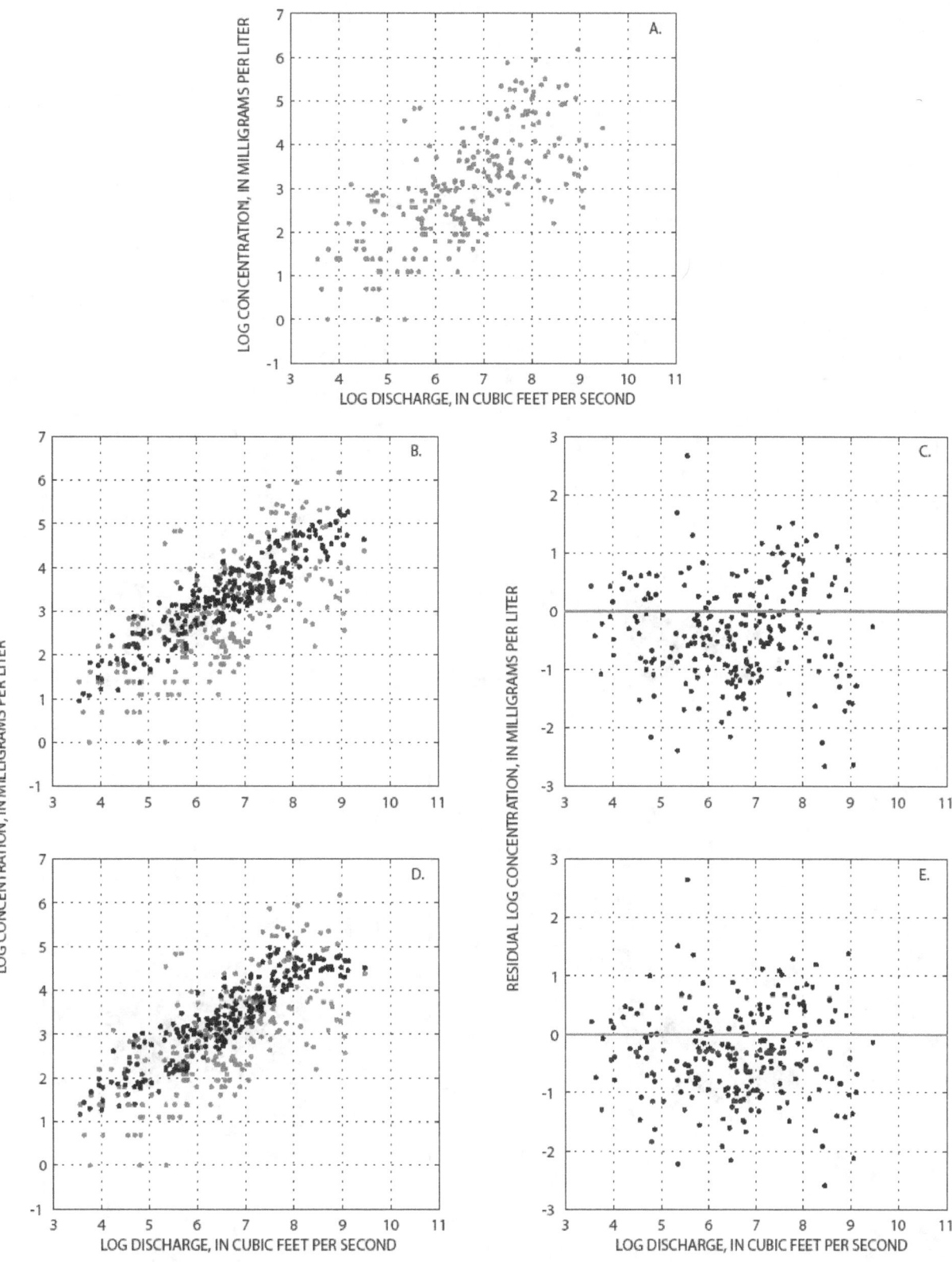

Figure 1–30.   Suspended sediment at Pamunkey River near Hanover, Virginia (USGS Station ID 01673000), showing the *(A)* observed concentration (red dots) versus discharge relation, *(B)* observed (red dots) and ESTIMATOR-predicted (black dots) concentration versus discharge relation, *(C)* residual (observed minus predicted) plot for ESTIMATOR predictions, *(D)* observed (red dots) and WRTDS-predicted (black dots) concentration versus discharge relation, and *(E)* residual (observed minus predicted) plot for WRTDS predictions.

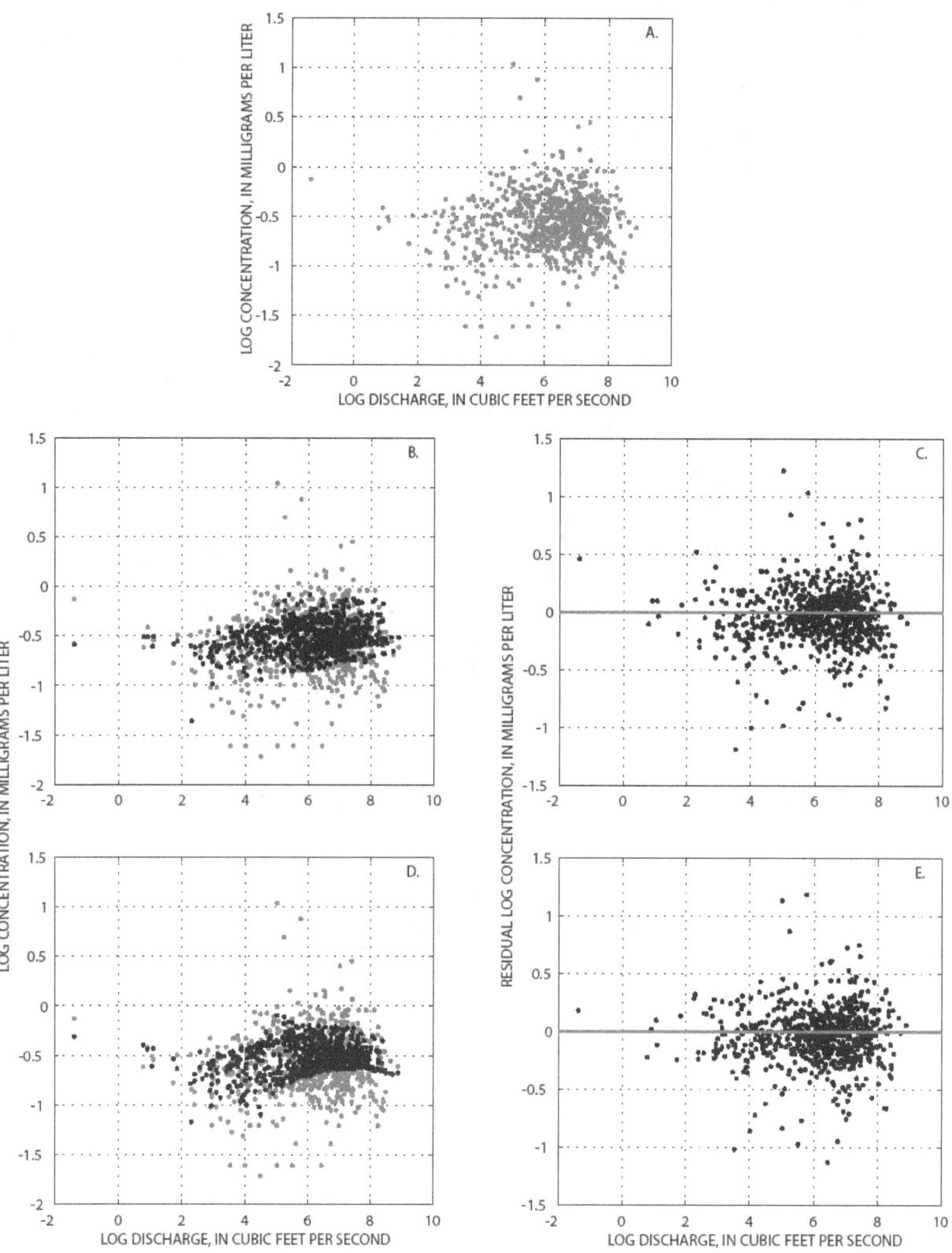

Figure 1–31.   Total nitrogen at Mattaponi River near Beulahville, Virginia (USGS Station ID 01674500), showing the *(A)* observed concentration (red dots) versus discharge relation, *(B)* observed (red dots) and ESTIMATOR-predicted (black dots) concentration versus discharge relation, *(C)* residual (observed minus predicted) plot for ESTIMATOR predictions, *(D)* observed (red dots) and WRTDS-predicted (black dots) concentration versus discharge relation, and *(E)* residual (observed minus predicted) plot for WRTDS predictions.

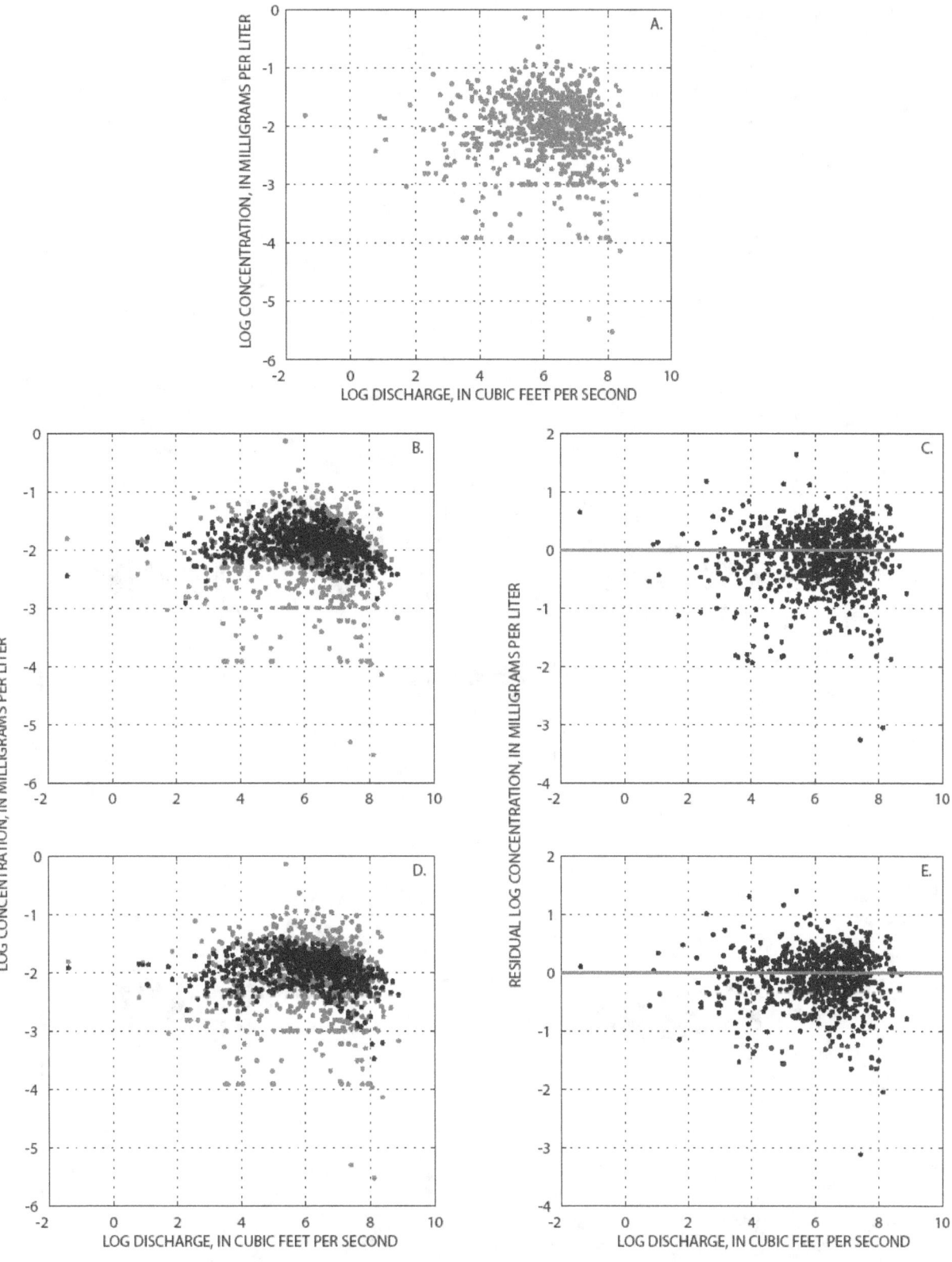

Figure 1–32. Nitrate at Mattaponi River near Beulahville, Virginia (USGS Station ID 01674500), showing the *(A)* observed concentration (red dots) versus discharge relation, *(B)* observed (red dots) and ESTIMATOR-predicted (black dots) concentration versus discharge relation, *(C)* residual (observed minus predicted) plot for ESTIMATOR predictions, *(D)* observed (red dots) and WRTDS-predicted (black dots) concentration versus discharge relation, and *(E)* residual (observed minus predicted) plot for WRTDS predictions.

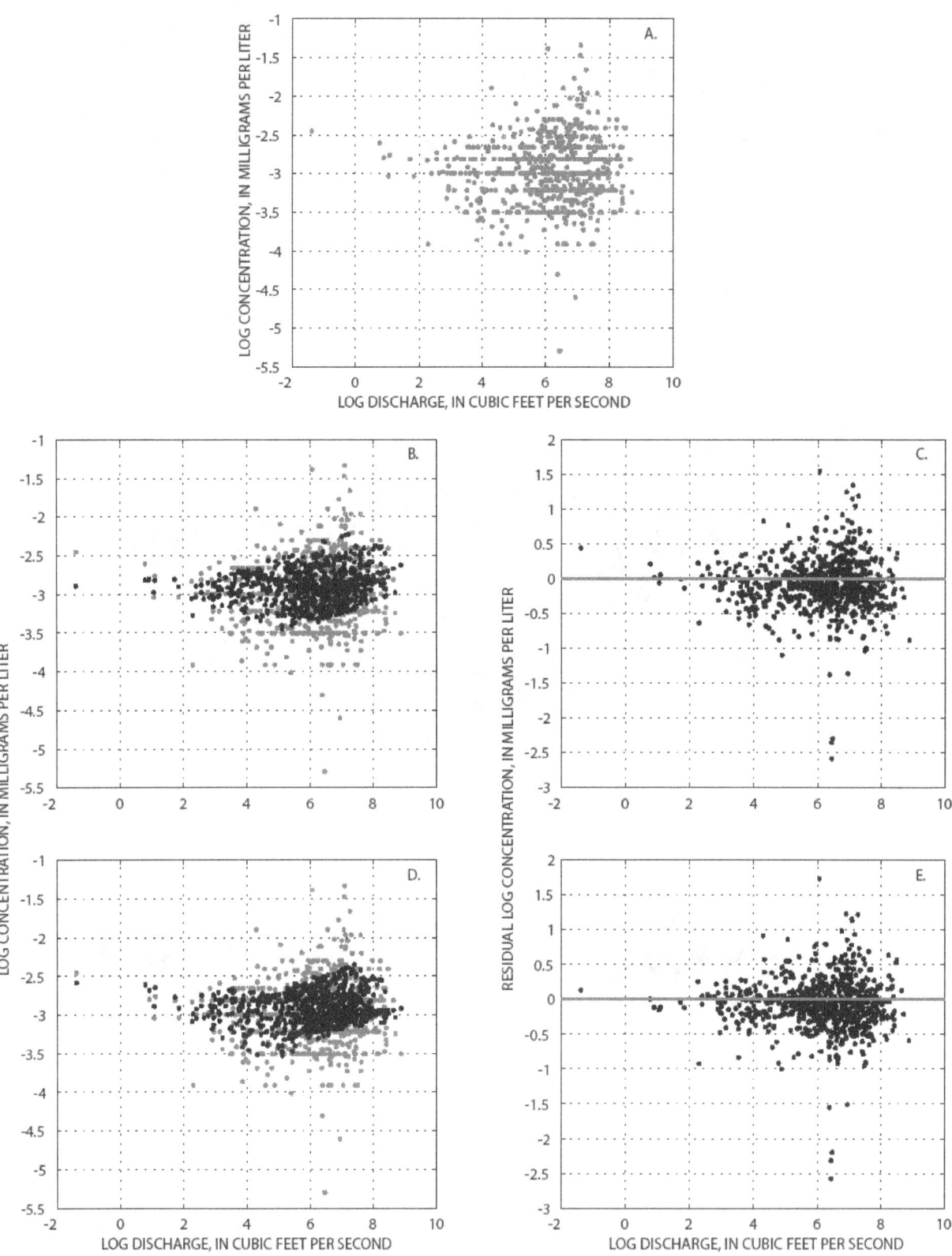

Figure 1–33.   Total phosphorus at Mattaponi River near Beulahville, Virginia (USGS Station ID 01674500), showing the *(A)* observed concentration (red dots) versus discharge relation, *(B)* observed (red dots) and ESTIMATOR-predicted (black dots) concentration versus discharge relation, *(C)* residual (observed minus predicted) plot for ESTIMATOR predictions, *(D)* observed (red dots) and WRTDS-predicted (black dots) concentration versus discharge relation, and *(E)* residual (observed minus predicted) plot for WRTDS predictions.

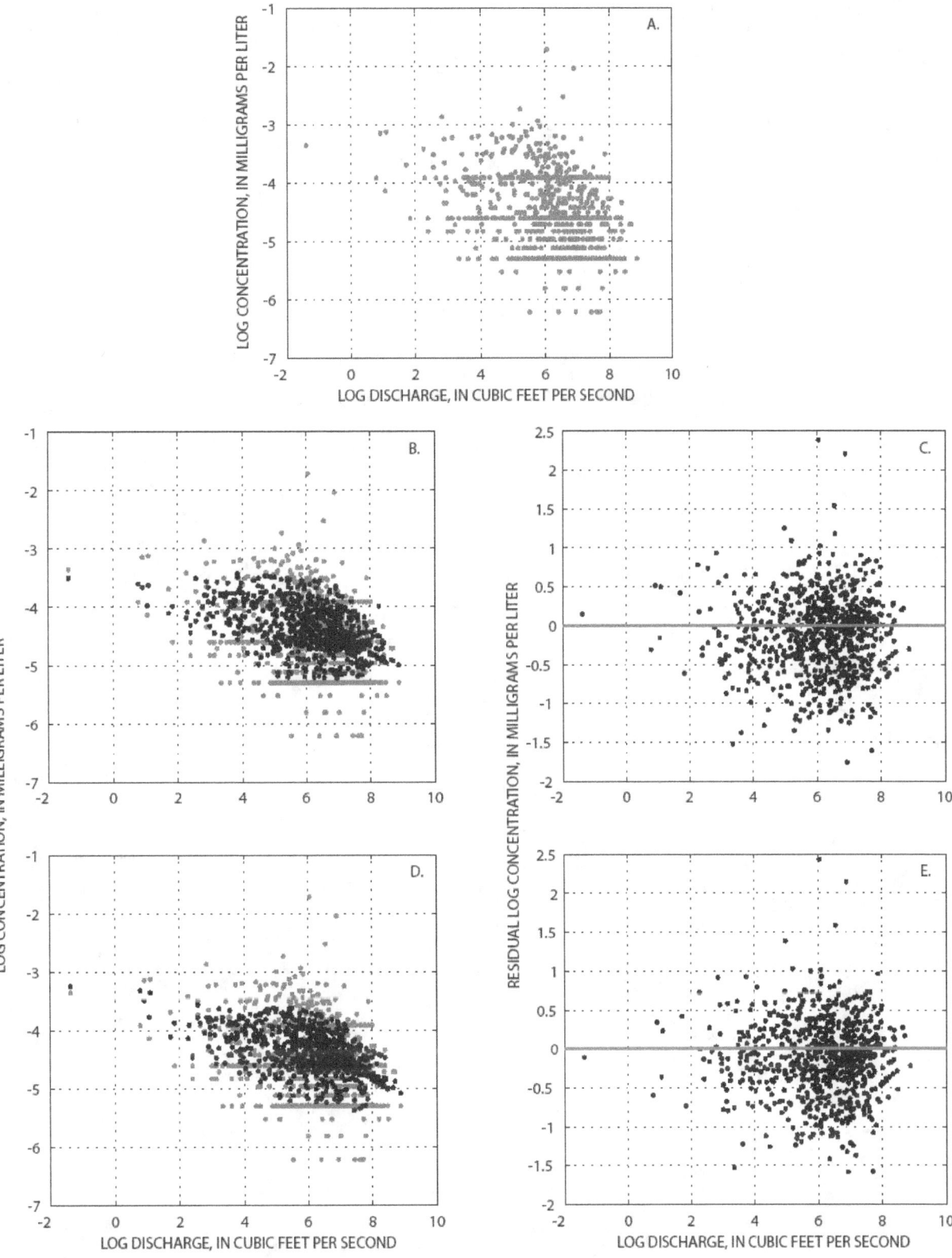

Figure 1–34.   Orthophosphorus at Mattaponi River near Beulahville, Virginia (USGS Station ID 01674500), showing the *(A)* observed concentration (red dots) versus discharge relation, *(B)* observed (red dots) and ESTIMATOR-predicted (black dots) concentration versus discharge relation, *(C)* residual (observed minus predicted) plot for ESTIMATOR predictions, *(D)* observed (red dots) and WRTDS-predicted (black dots) concentration versus discharge relation, and *(E)* residual (observed minus predicted) plot for WRTDS predictions.

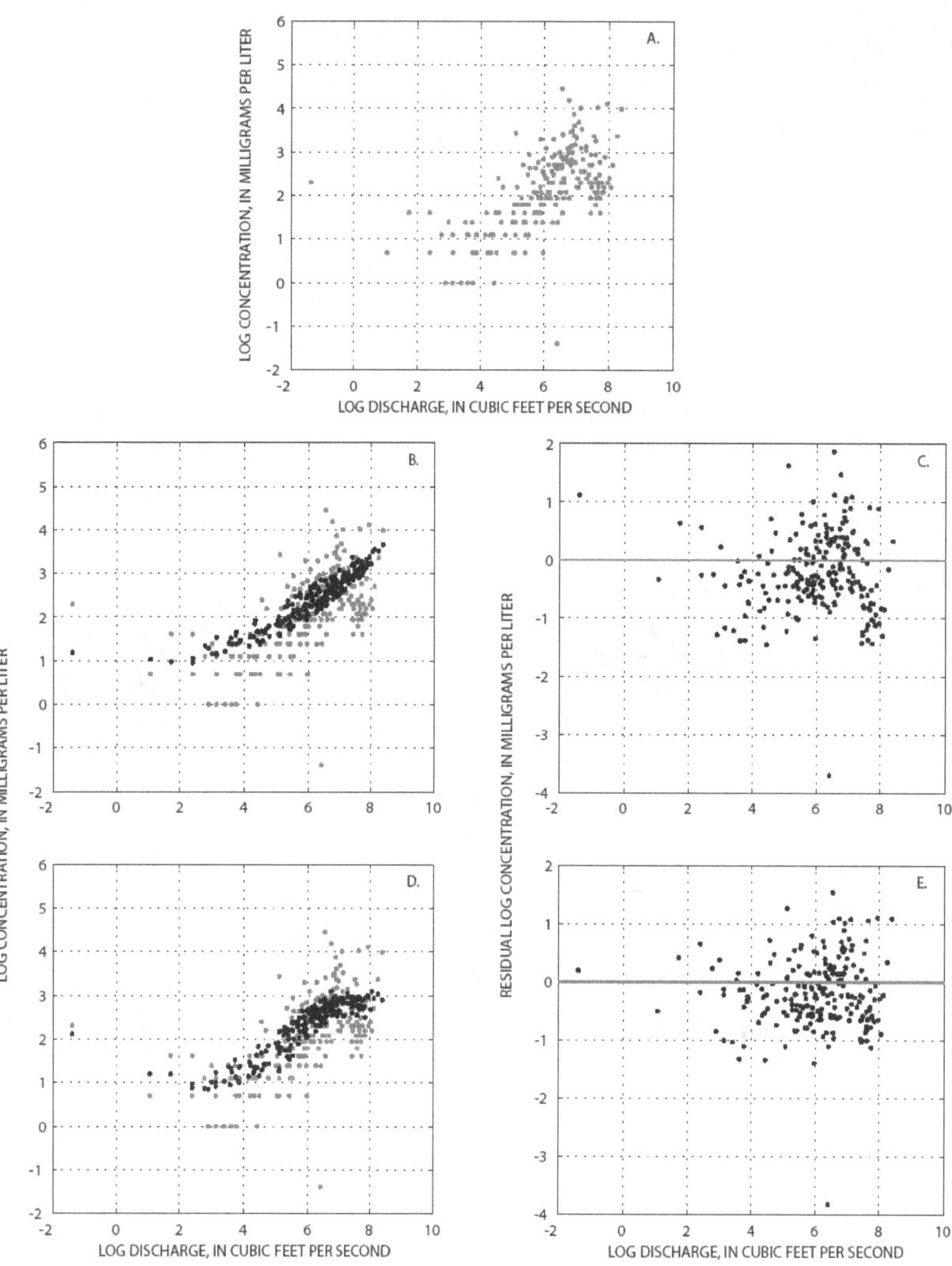

Figure 1–35.   Suspended sediment at Mattaponi River near Beulahville, Virginia (USGS Station ID 01674500), showing the *(A)* observed concentration (red dots) versus discharge relation, *(B)* observed (red dots) and ESTIMATOR-predicted (black dots) concentration versus discharge relation, *(C)* residual (observed minus predicted) plot for ESTIMATOR predictions, *(D)* observed (red dots) and WRTDS-predicted (black dots) concentration versus discharge relation, and *(E)* residual (observed minus predicted) plot for WRTDS predictions.

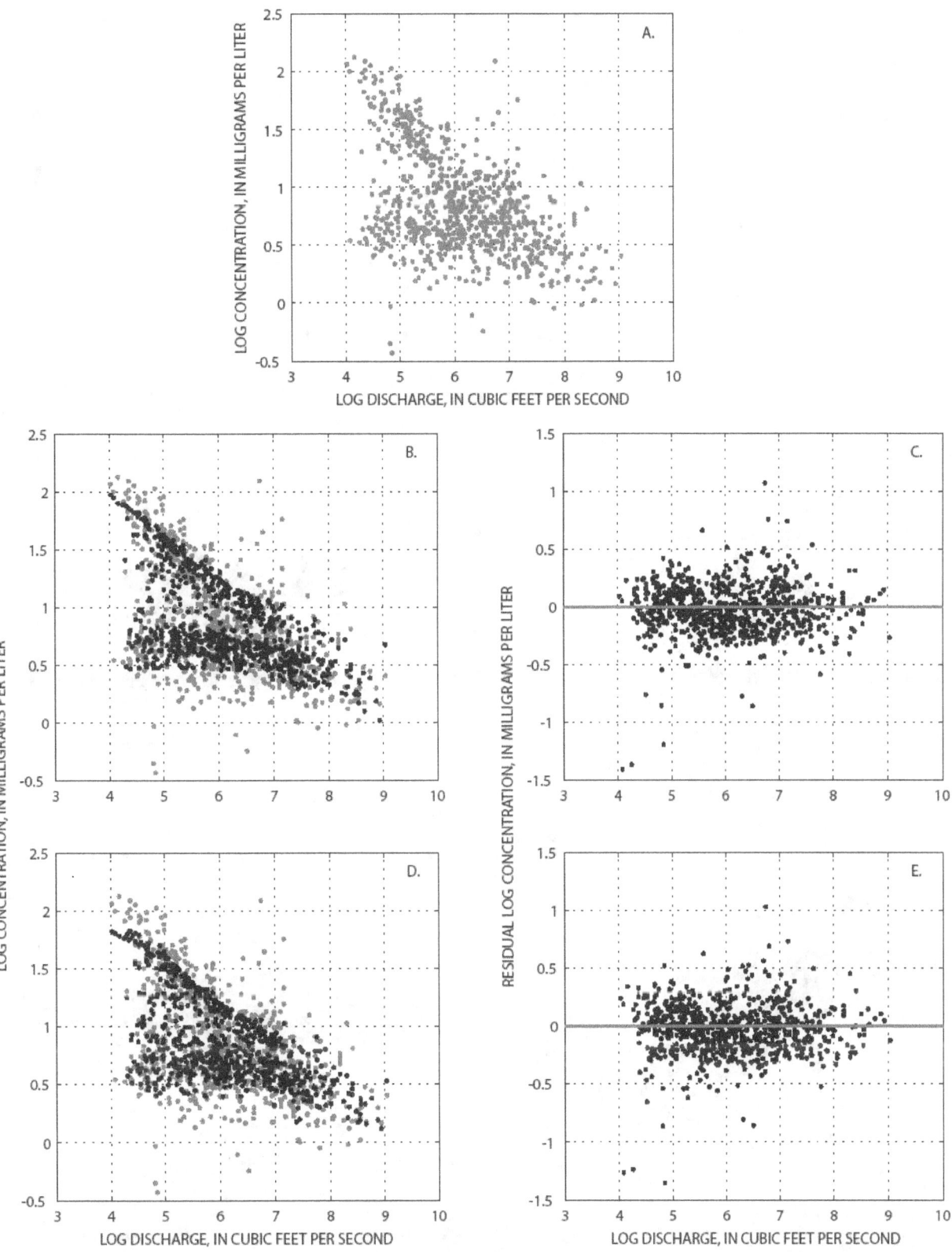

Figure 1–36. Total nitrogen at Patuxent River near Bowie, Maryland (USGS Station ID 01594440), showing the *(A)* observed concentration (red dots) versus discharge relation, *(B)* observed (red dots) and ESTIMATOR-predicted (black dots) concentration versus discharge relation, *(C)* residual (observed minus predicted) plot for ESTIMATOR predictions, *(D)* observed (red dots) and WRTDS-predicted (black dots) concentration versus discharge relation, and *(E)* residual (observed minus predicted) plot for WRTDS predictions.

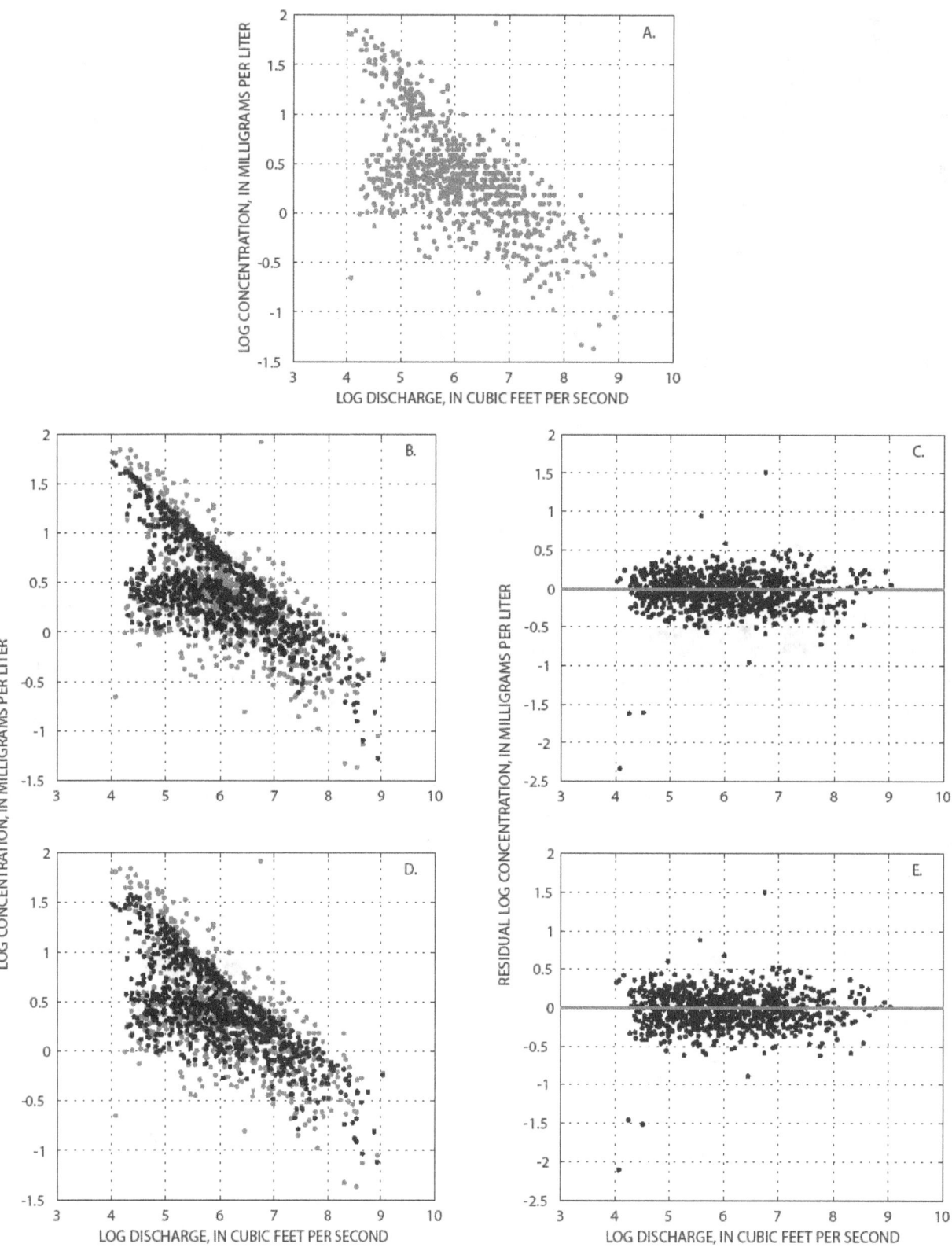

Figure 1–37.   Nitrate at Patuxent River near Bowie, Maryland (USGS Station ID 01594440), showing the *(A)* observed concentration (red dots) versus discharge relation, *(B)* observed (red dots) and ESTIMATOR-predicted (black dots) concentration versus discharge relation, *(C)* residual (observed minus predicted) plot for ESTIMATOR predictions, *(D)* observed (red dots) and WRTDS-predicted (black dots) concentration versus discharge relation, and *(E)* residual (observed minus predicted) plot for WRTDS predictions.

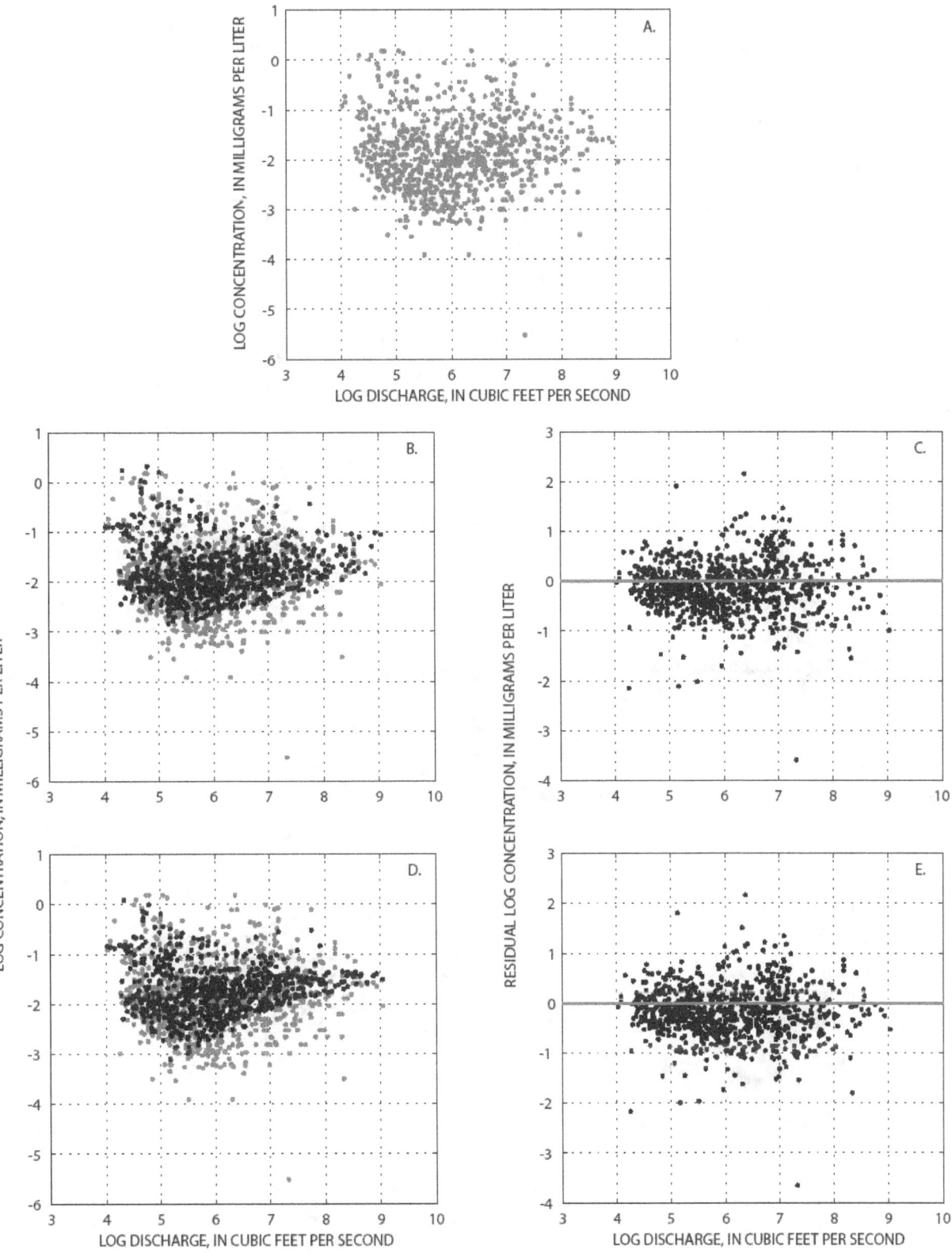

Figure 1–38.   Total phosphorus at Patuxent River near Bowie, Maryland (USGS Station ID 01594440), showing the *(A)* observed concentration (red dots) versus discharge relation, *(B)* observed (red dots) and ESTIMATOR-predicted (black dots) concentration versus discharge relation, *(C)* residual (observed minus predicted) plot for ESTIMATOR predictions, *(D)* observed (red dots) and WRTDS-predicted (black dots) concentration versus discharge relation, and *(E)* residual (observed minus predicted) plot for WRTDS predictions.

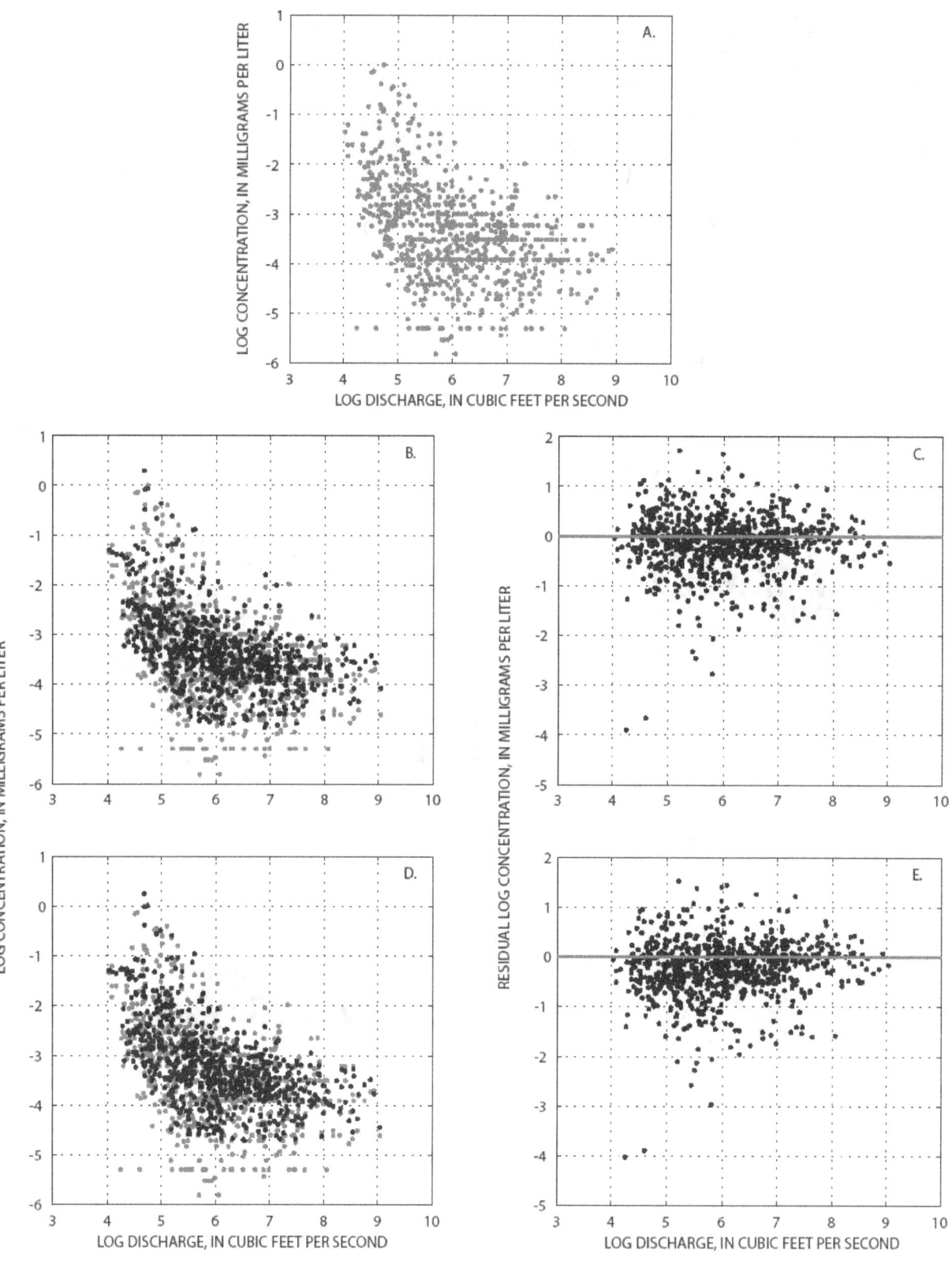

Figure 1– 39.   Orthophosphorus at Patuxent River near Bowie, Maryland (USGS Station ID 01594440), showing the *(A)* observed concentration (red dots) versus discharge relation, *(B)* observed (red dots) and ESTIMATOR-predicted (black dots) concentration versus discharge relation, *(C)* residual (observed minus predicted) plot for ESTIMATOR predictions, *(D)* observed (red dots) and WRTDS-predicted (black dots) concentration versus discharge relation, and *(E)* residual (observed minus predicted) plot for WRTDS predictions.

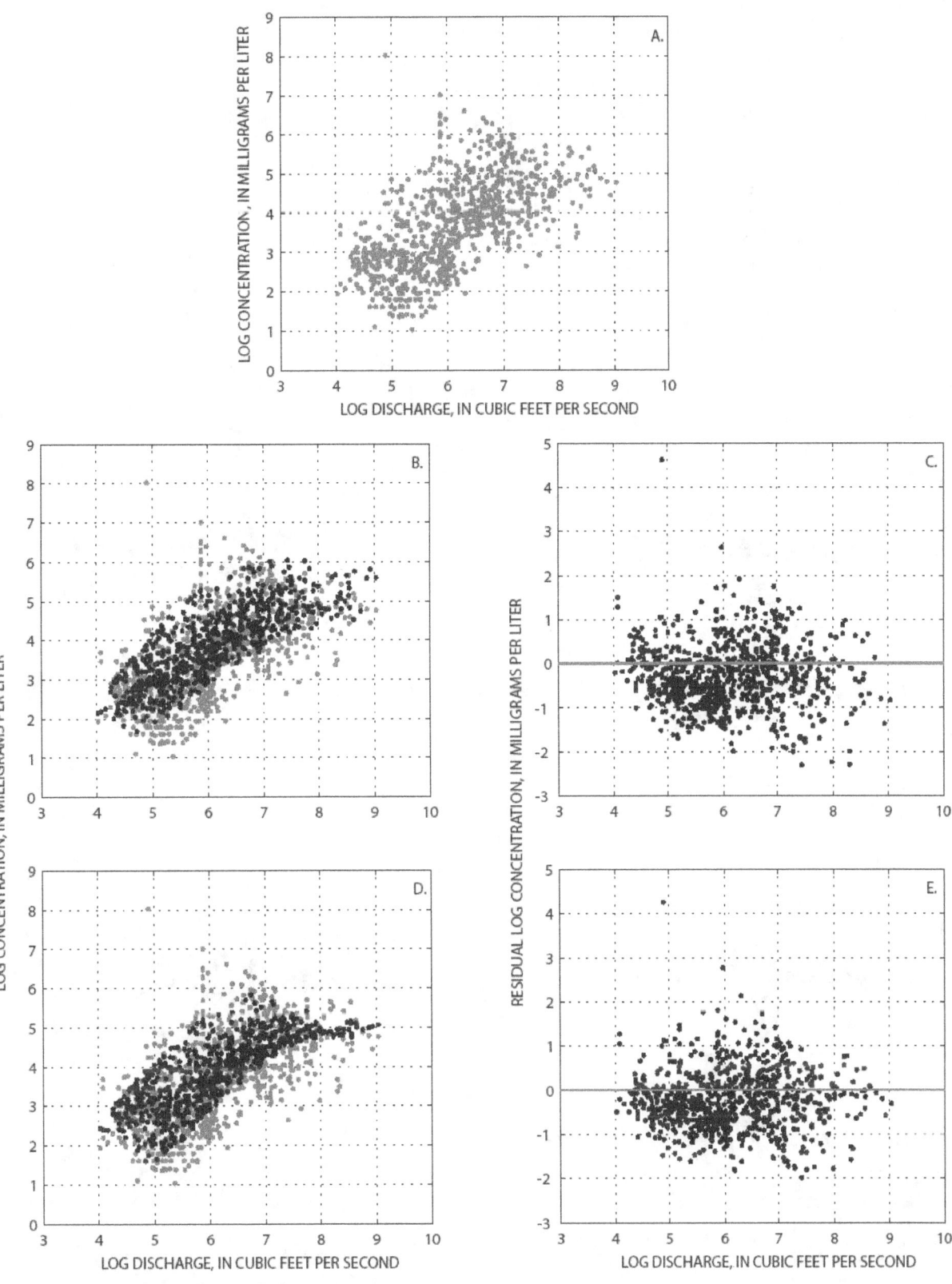

Figure 1–40. Suspended sediment at Patuxent River near Bowie, Maryland (USGS Station ID 01594440), showing the *(A)* observed concentration (red dots) versus discharge relation, *(B)* observed (red dots) and ESTIMATOR-predicted (black dots) concentration versus discharge relation, *(C)* residual (observed minus predicted) plot for ESTIMATOR predictions, *(D)* observed (red dots) and WRTDS-predicted (black dots) concentration versus discharge relation, and *(E)* residual (observed minus predicted) plot for WRTDS predictions.

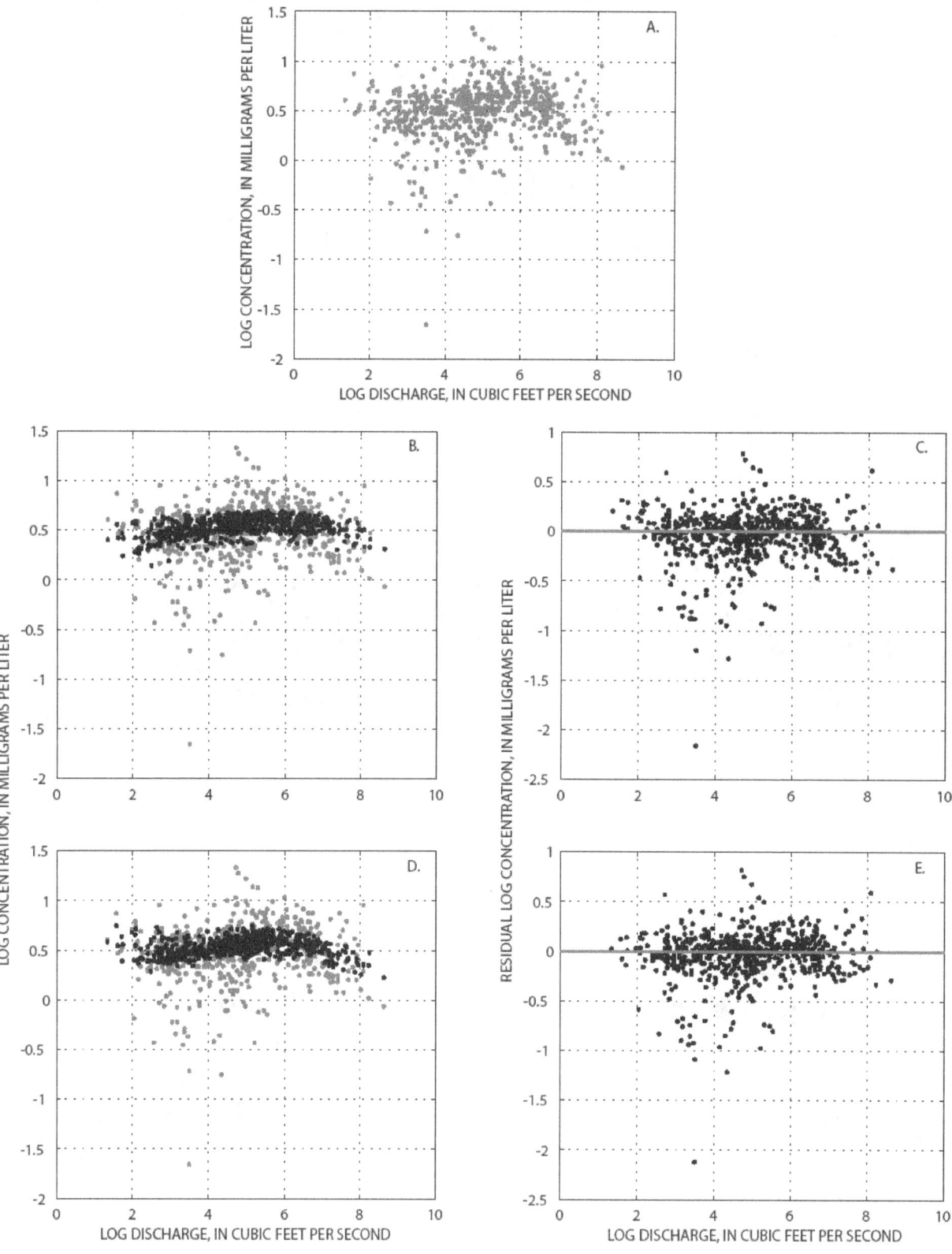

Figure 1–41.   Total nitrogen at Choptank River near Greensboro, Maryland (USGS Station ID 01491000), showing the
(A) observed concentration (red dots) versus discharge relation, *(B)* observed (red dots) and ESTIMATOR-predicted
(black dots) concentration versus discharge relation, *(C)* residual (observed minus predicted) plot for ESTIMATOR
predictions, *(D)* observed (red dots) and WRTDS-predicted (black dots) concentration versus discharge relation, and
(E) residual (observed minus predicted) plot for WRTDS predictions.

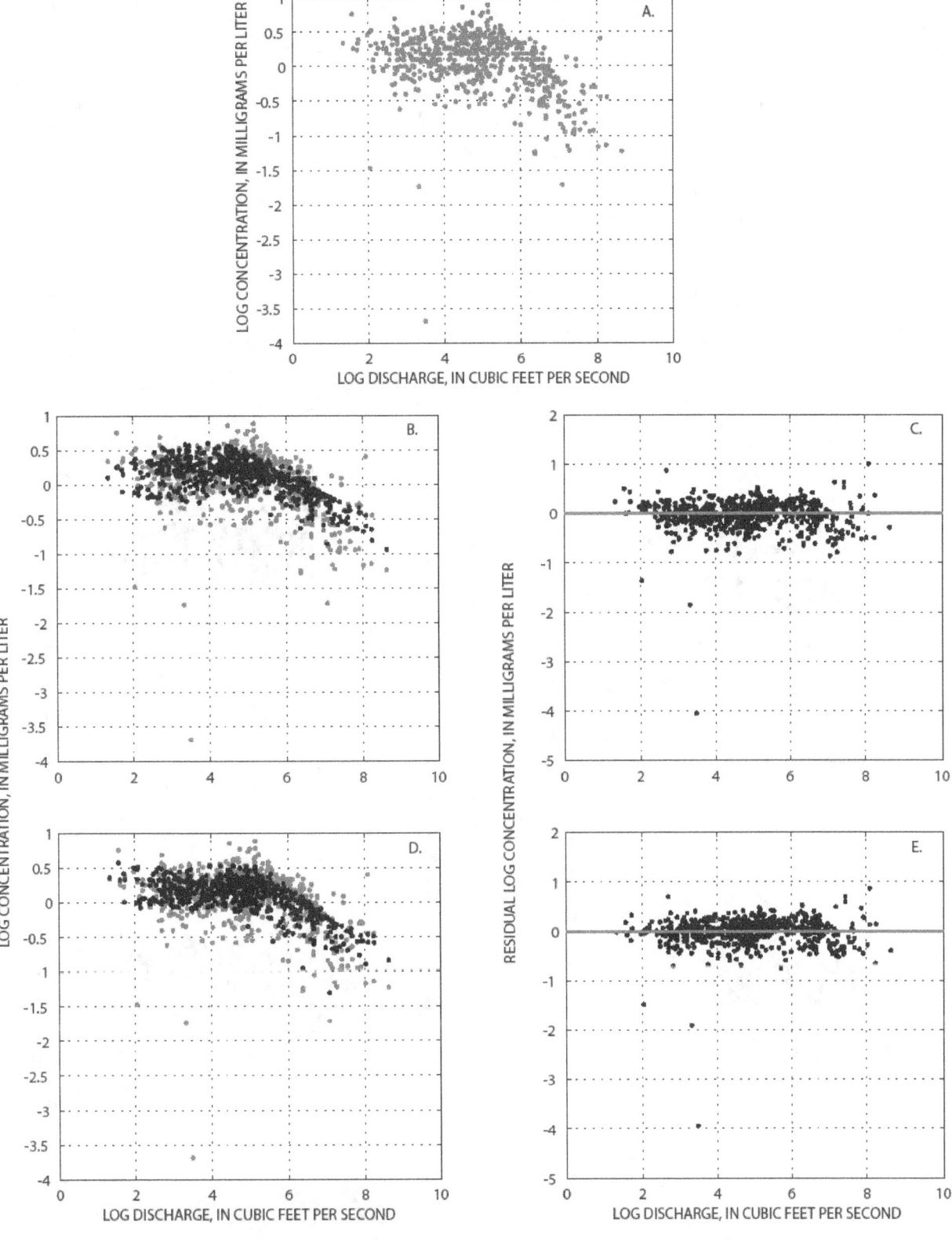

Figure 1–42. Nitrate at Choptank River near Greensboro, Maryland (USGS Station ID 01491000), showing the *(A)* observed concentration (red dots) versus discharge relation, *(B)* observed (red dots) and ESTIMATOR-predicted (black dots) concentration versus discharge relation, *(C)* residual (observed minus predicted) plot for ESTIMATOR predictions, *(D)* observed (red dots) and WRTDS-predicted (black dots) concentration versus discharge relation, and *(E)* residual (observed minus predicted) plot for WRTDS predictions.

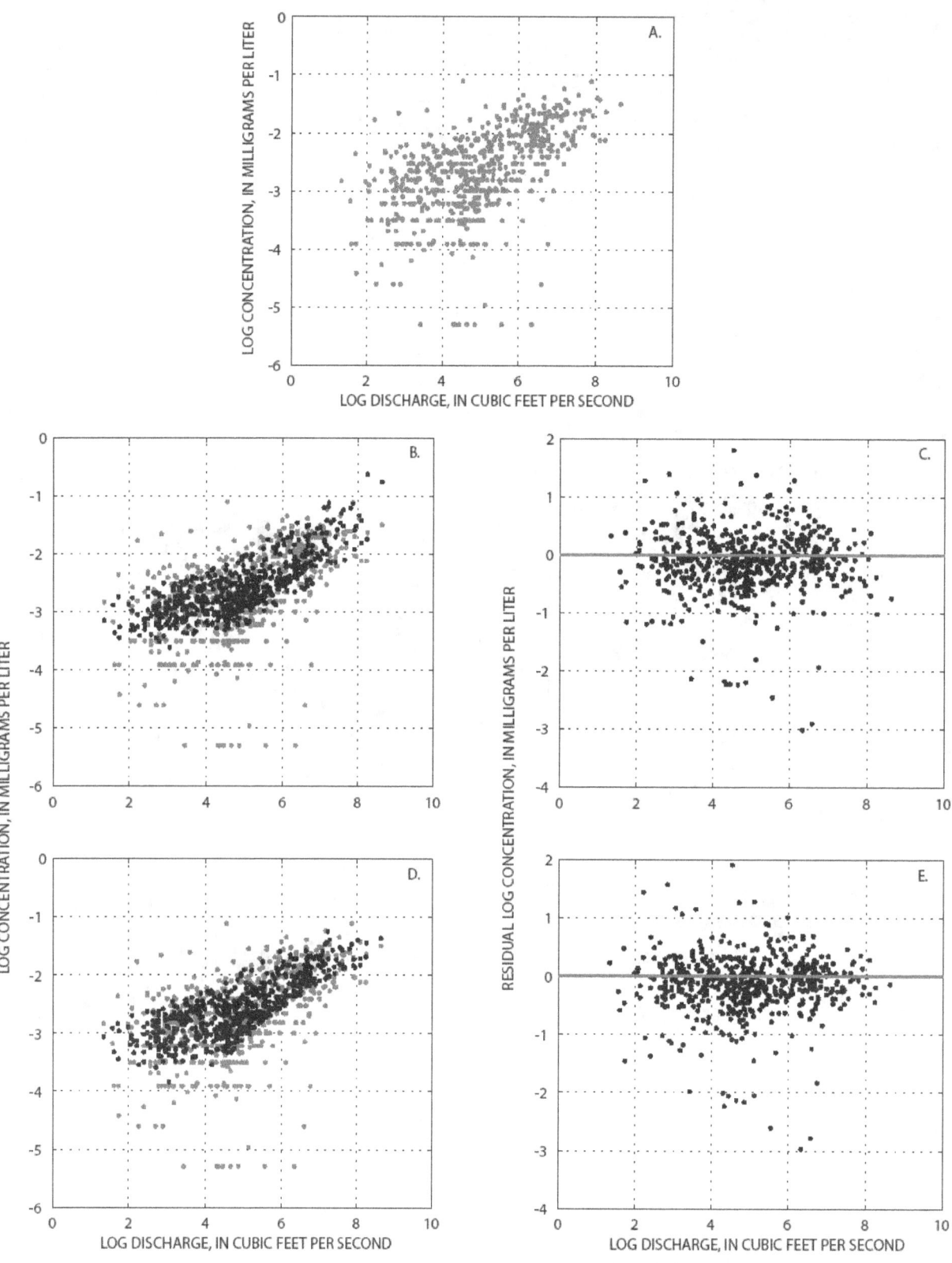

Figure 1–43.   Total phosphorus at Choptank River near Greensboro, Maryland (USGS Station ID 01491000), showing the *(A)* observed concentration (red dots) versus discharge relation, *(B)* observed (red dots) and ESTIMATOR-predicted (black dots) concentration versus discharge relation, *(C)* residual (observed minus predicted) plot for ESTIMATOR predictions, *(D)* observed (red dots) and WRTDS-predicted (black dots) concentration versus discharge relation, and *(E)* residual (observed minus predicted) plot for WRTDS predictions.

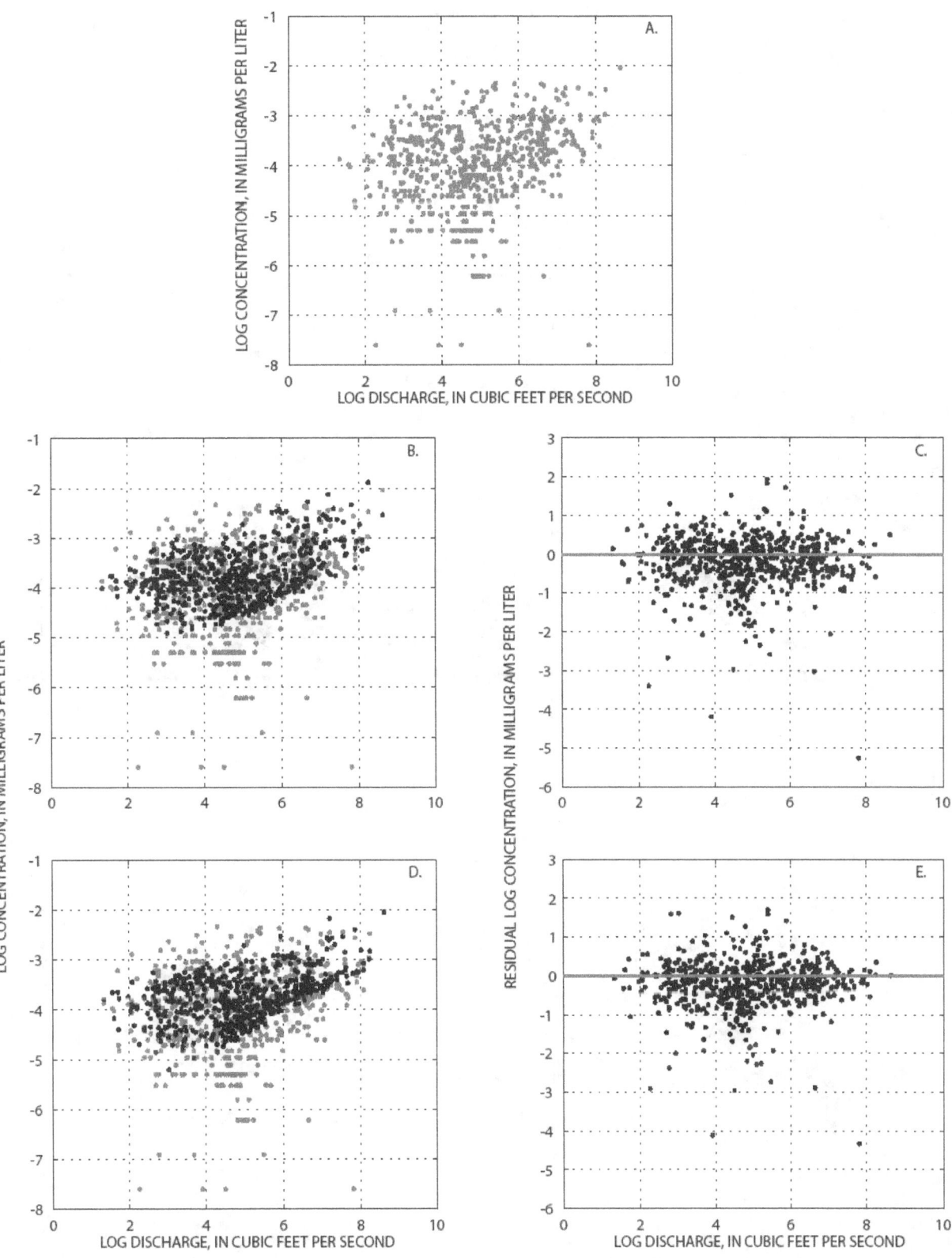

Figure 1–44. Orthophosphorus at Choptank River near Greensboro, Maryland (USGS Station ID 01491000), showing the *(A)* observed concentration (red dots) versus discharge relation, *(B)* observed (red dots) and ESTIMATOR-predicted (black dots) concentration versus discharge relation, *(C)* residual (observed minus predicted) plot for ESTIMATOR predictions, *(D)* observed (red dots) and WRTDS-predicted (black dots) concentration versus discharge relation, and *(E)* residual (observed minus predicted) plot for WRTDS predictions.

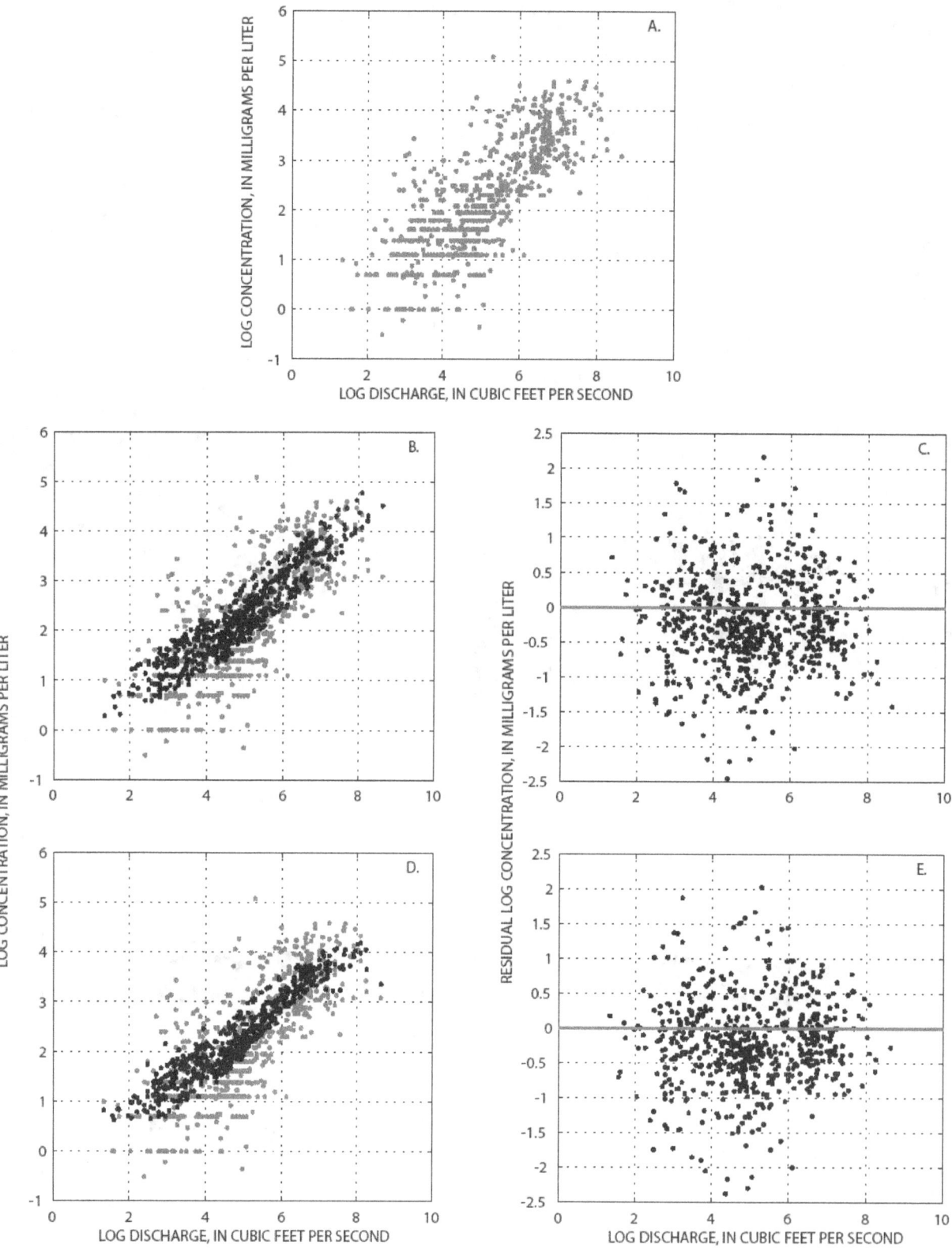

Figure 1–45. Suspended sediment at Choptank River near Greensboro, Maryland (USGS Station ID 01491000), showing the *(A)* observed concentration (red dots) versus discharge relation, *(B)* observed (red dots) and ESTIMATOR-predicted (black dots) concentration versus discharge relation, *(C)* residual (observed minus predicted) plot for ESTIMATOR predictions, *(D)* observed (red dots) and WRTDS-predicted (black dots) concentration versus discharge relation, and *(E)* residual (observed minus predicted) plot for WRTDS predictions.

Appendix 2. WRTDS-derived annual and flow-normalized annual fluxes for total nitrogen, nitrate, total phosphorus, orthophosphorus, and suspended sediment at the nine Chesapeake Bay River Input Monitoring stations

_____

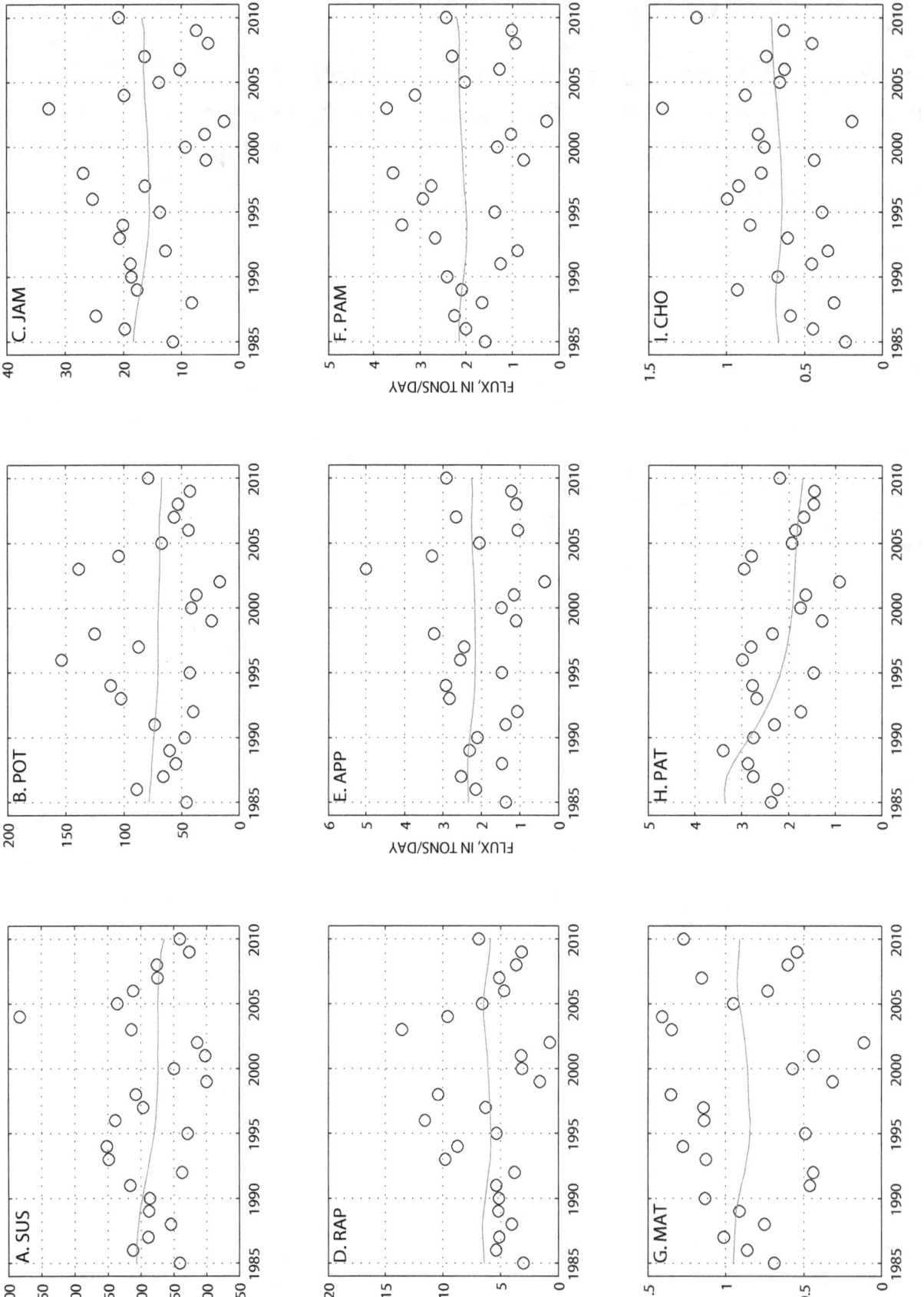

Figure 2–1. WRTDS-derived total nitrogen annual fluxes (black circles) and flow-normalized fluxes (red line) at the (*A*) Susquehanna, (*B*) Potomac, (*C*) James, (*D*) Rappahannock, (*E*) Appomattox, (*F*) Pamunkey, (*G*) Mattaponi, (*H*) Patuxent, (*I*) Choptank River Input Monitoring stations.

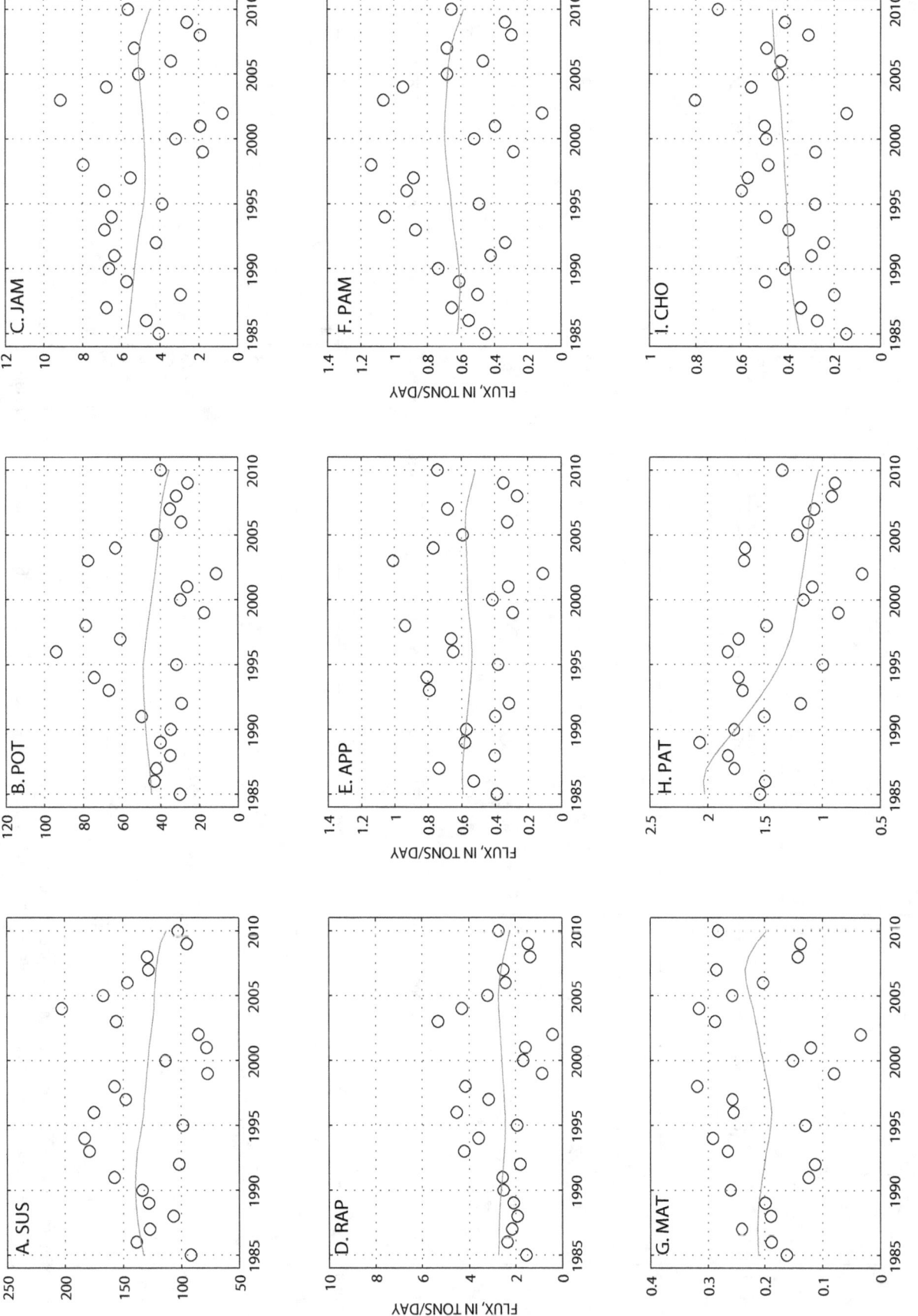

Figure 2–2. WRTDS-derived nitrate annual fluxes (black circles) and flow-normalized fluxes (red line) at the *(A)* Susquehanna, *(B)* Potomac, *(C)* James, *(D)* Rappahannock, *(E)* Appomattox, *(F)* Pamunkey, *(G)* Mattaponi, *(H)* Patuxent, *(I)* Choptank River Input Monitoring stations.

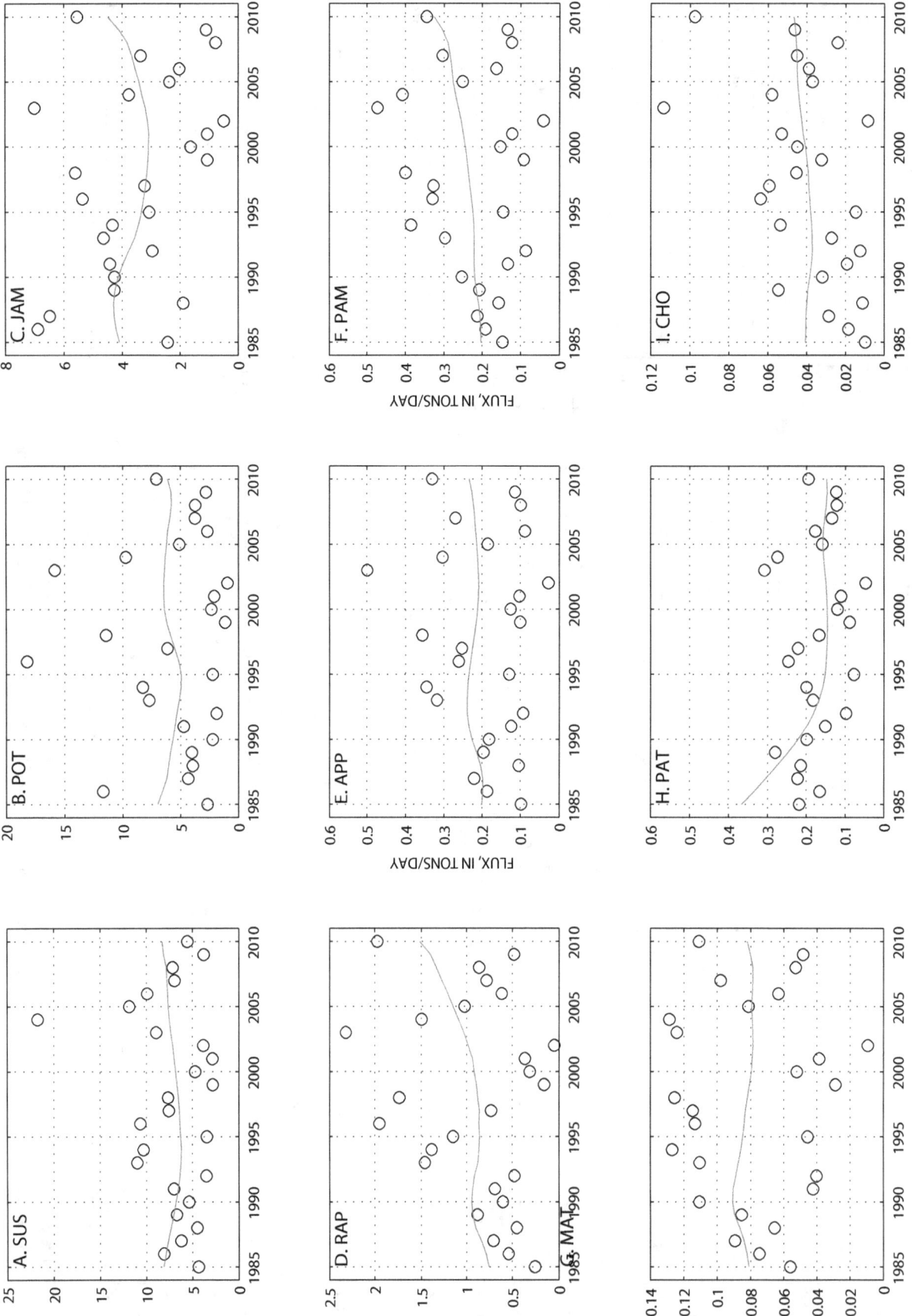

Figure 2–3.   WRTDS-derived total phosphorus annual fluxes (black circles) and flow-normalized fluxes (red line) at the *(A)* Susquehanna, *(B)* Potomac, *(C)* James, *(D)* Rappahannock, *(E)* Appomattox, *(F)* Pamunkey, *(G)* Mattaponi, *(H)* Patuxent, *(I)* Choptank River Input Monitoring stations.

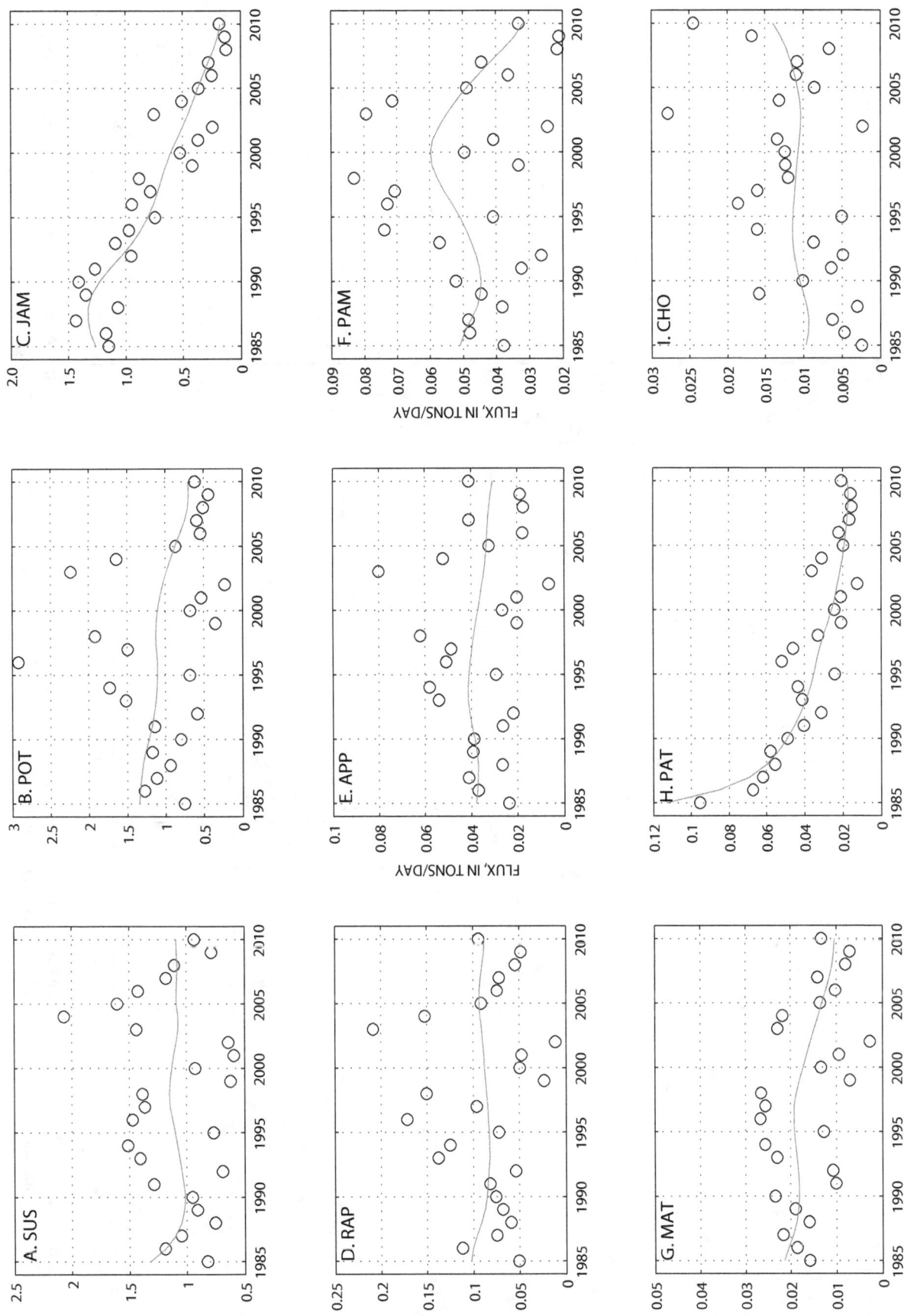

Figure 2–4. WRTDS-derived orthophosphorus annual fluxes (black circles) and flow-normalized fluxes (red line) at the *(A)* Susquehanna, *(B)* Potomac, *(C)* James, *(D)* Rappahannock, *(E)* Appomattox, *(F)* Pamunkey, *(G)* Mattaponi, *(H)* Patuxent, *(I)* Choptank River Input Monitoring stations.

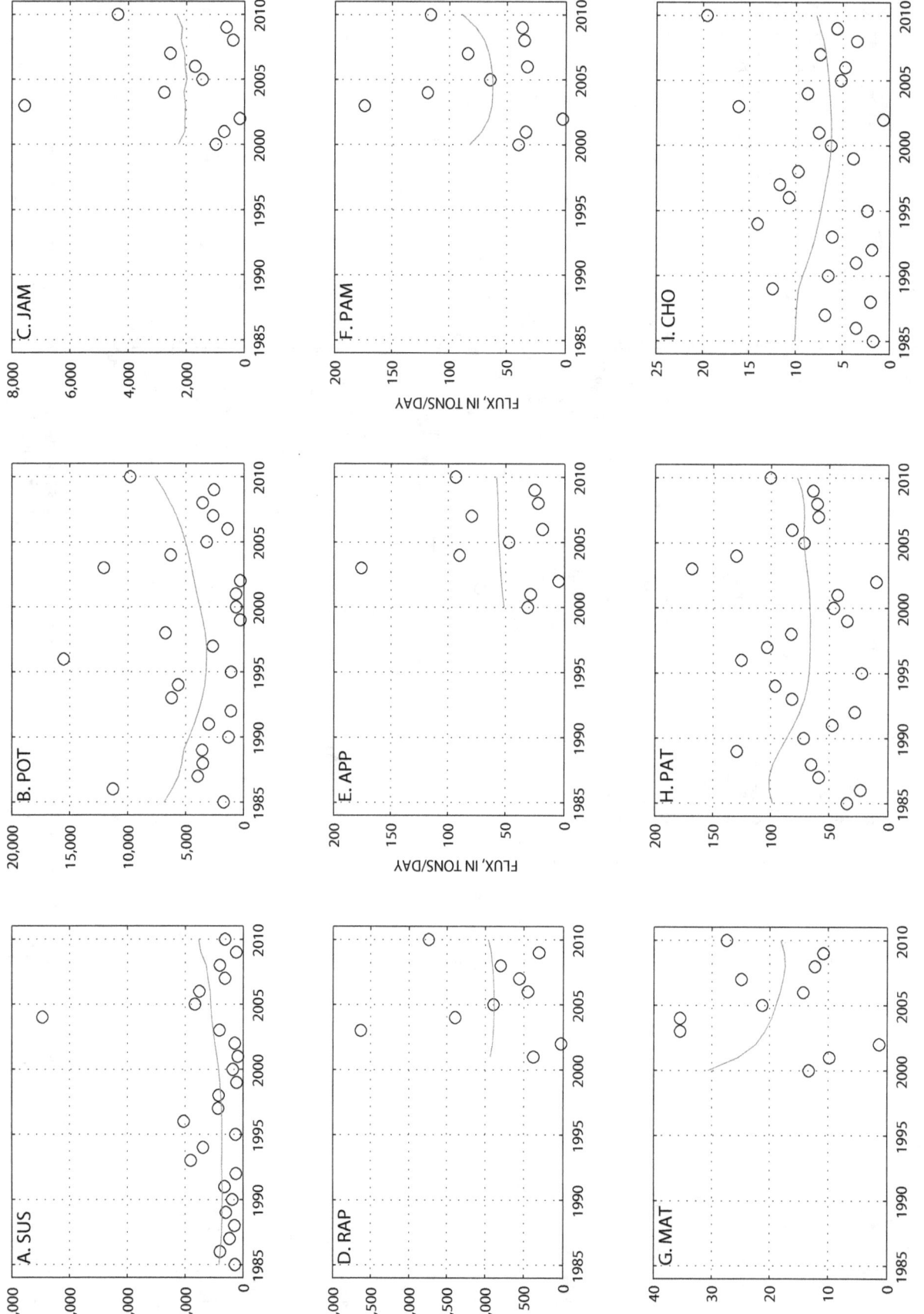

Figure 2–5. WRTDS-derived suspended sediment annual fluxes (black circles) and flow-normalized fluxes (red line) at the (A) Susquehanna, (B) Potomac, (C) James, (D) Rappahannock, (E) Appomattox, (F) Pamunkey, (G) Mattaponi, (H) Patuxent, (I) Choptank River Input Monitoring stations.

Appendix 3. Rates of change in WRTDS flow-normalized flux for total nitrogen, nitrate, total phosphorus, orthophosphorus, and suspended sediment at the nine Chesapeake Bay River Input Monitoring stations

Table 3–1. Total nitrogen trend slopes for WRTDS flow-normalized fluxes at the nine River Input Monitoring (RIM) stations for the time periods 1985 to 2010 and 2001 to 2010.

[ton/d/yr, tons per day per year; %/yr, percent per year]

| RIM station | WRTDS flow-normalized flux | | | |
| | 1985 to 2010 | | 2001 to 2010 | |
	Slope × 10⁻² (ton/d/yr)	Slope (%/yr)	Slope × 10⁻² (ton/d/yr)	Slope (%/yr)
Susquehanna	−171.79	−0.8	−111.98	−0.6
Potomac	−44.90	−0.6	−27.95	−0.4
James	−6.03	−0.3	11.44	0.7
Rappahannock	−2.19	−0.3	−2.95	−0.5
Appomattox	−0.32	−0.1	0.79	0.4
Pamunkey	0.24	0.1	1.14	0.5
Mattaponi	−0.16	−0.2	0.46	0.5
Patuxent	−6.62	−2.0	−2.24	−1.2
Choptank	0.20	0.3	0.55	0.8

Table 3–2. Nitrate trend slopes for WRTDS flow-normalized fluxes at the nine River Input Monitoring (RIM) stations for the time periods 1985 to 2010 and 2001 to 2010.

[ton/d/yr, tons per day per year; %/yr, percent per year]

| RIM station | WRTDS flow-normalized flux | | | |
| | 1985 to 2010 | | 2001 to 2010 | |
	Slope × 10⁻² (ton/d/yr)	Slope (%/yr)	Slope × 10⁻² (ton/d/yr)	Slope (%/yr)
Susquehanna	−79.13	−0.6	−162.80	−1.3
Potomac	−37.47	−0.8	−91.98	−2.1
James	−4.71	−0.8	−3.96	−0.8
Rappahannock	−2.04	−0.7	−4.31	−1.6
Appomattox	−0.31	−0.5	−0.52	−0.9
Pamunkey	−0.16	−0.3	−1.28	−1.8
Mattaponi	−0.05	−0.3	−0.11	−0.5
Patuxent	−3.98	−2.0	−1.75	−1.5
Choptank	0.46	1.3	0.46	1.1

Table 3–3.   Total phosphorus trend slopes for WRTDS flow-normalized fluxes at the nine River Input Monitoring stations for the time periods 1985 to 2010 and 2001 to 2010.

[ton/d/yr, tons per day per year; %/yr, percent per year]

| RIM station | WRTDS flow-normalized flux | | | |
| | 1985 to 2010 | | 2001 to 2010 | |
	Slope × 10⁻³ (ton/d/yr)	Slope (%/yr)	Slope × 10⁻³ (ton/d/yr)	Slope (%/yr)
Susquehanna	6.10	0.1	143.33	2.0
Potomac	−34.71	−0.5	−36.02	−0.6
James	15.43	0.4	157.02	5.1
Rappahannock	30.30	4.0	64.50	6.9
Appomattox	1.30	0.6	2.68	1.3
Pamunkey	4.92	2.4	8.62	3.5
Mattaponi	0.04	0.0	0.32	0.4
Patuxent	−8.75	−2.4	0.04	0.0
Choptank	0.24	0.6	0.54	1.3

Table 3–4.   Orthophosphorus trend slopes for WRTDS flow-normalized fluxes at the nine River Input Monitoring stations for the time periods 1985 to 2010 and 2001 to 2010.

[ton/y/yr, tons per day per year; %/yr, percent per year]

| RIM station | WRTDS flow-normalized flux | | | |
| | 1985 to 2010 | | 2001 to 2010 | |
	Slope × 10⁻³ (ton/d/yr)	Slope (%/yr)	Slope × 10⁻³ (ton/d/yr)	Slope (%/yr)
Susquehanna	−9.00	−0.7	−2.72	−0.2
Potomac	−25.76	−1.9	−41.47	−3.9
James	−43.44	−3.4	−42.03	−7.6
Rappahannock	−0.58	−0.6	−0.04	0.0
Appomattox	−0.29	−0.8	−0.61	−1.7
Pamunkey	−0.77	−1.5	−3.04	−5.1
Mattaponi	−0.44	−2.0	−0.68	−4.1
Patuxent	−3.82	−3.4	−0.74	−3.1
Choptank	0.17	1.8	0.38	3.6

Table 3–5.   Suspended sediment trend slopes for flow-normalized at the nine River Input Monitoring stations for the time periods 1985 to 2010 and 2001 to 2010.

[ton/d/yr, tons per day per year; %/yr, percent per year; NA, not available]

| RIM station | WRTDS flow-normalized yield | | | |
| | 1985 to 2010 | | 2001 to 2010 | |
	Slope (ton/d/yr)	Slope (%/yr)	Slope (ton/d/yr)	Slope (%/yr)
Susquehanna	143.11	3.3	356.21	7.9
Potomac	33.16	0.5	400.12	9.9
James	NA	NA	32.39	1.6
Rappahannock	NA	NA	2.48	0.3
Appomattox	NA	NA	0.59	1.1
Pamunkey	NA	NA	3.68	2.6
Mattaponi	NA	NA	−0.81	−3.2
Patuxent	−0.82	−0.8	1.20	1.8
Choptank	−0.09	−0.9	0.18	2.9

Appendix 4. Time series of WRTDS annual and flow-normalized annual fluxes of total nitrogen, nitrate, total phosphorus, orthophosphorus, and suspended sediment at the nine River Input Monitoring stations in the Chesapeake Bay watershed

Table 4–1.   Total nitrogen annual flux and flow-normalized annual flux, in tons per day, for the nine Chesapeake Bay River Input Monitoring Stations obtained using WRTDS.

[SUS, Susquehanna River at Conowingo, Md.; POT, Potomac River at Chain Bridge at Washington, D.C.; JAM, James River at Cartersville, Va.; RAP, Rappahannock River near Fredericksburg, Va.; APP, Appomattox River at Matoaca, Va.; PAM, Pamunkey River near Hanover, Va.; MAT, Mattaponi River near Beulahville, Va.; PAT, Patuxent River near Bowie, Md.; CHO, Choptank River near Greensboro, Md.; FN flux, flow-normalized flux]

Year	SUS Flux	SUS FN flux	POT Flux	POT FN flux	JAM Flux	JAM FN flux	RAP Flux	RAP FN flux	APP Flux	APP FN flux	PAM Flux	PAM FN flux	MAT Flux	MAT FN flux	PAT Flux	PAT FN flux	CHO Flux	CHO FN flux
1981	131.77	199.25	30.98	74.27	5.72	19.23	1.39	6.42	0.78	2.30	0.47	2.18	0.32	0.92	1.98	3.10	0.34	0.62
1982	190.46	202.72	62.34	76.50	17.15	18.89	4.75	6.50	1.97	2.31	1.63	2.16	0.64	0.94	2.34	3.18	0.49	0.63
1983	177.20	204.76	73.40	77.96	19.53	18.63	7.00	6.47	2.75	2.31	2.14	2.14	1.16	0.95	3.25	3.26	0.77	0.64
1984	279.39	205.62	124.82	78.58	28.30	18.48	11.65	6.47	3.52	2.32	3.75	2.14	1.67	0.95	4.15	3.32	0.93	0.65
1985	140.84	206.92	45.70	78.26	11.44	18.31	3.01	6.45	1.37	2.33	1.59	2.15	0.69	0.95	2.37	3.36	0.24	0.67
1986	212.21	206.66	88.96	77.51	19.78	18.12	5.39	6.49	2.14	2.34	2.00	2.15	0.86	0.95	2.24	3.37	0.45	0.68
1987	188.83	205.86	65.89	76.20	24.72	17.83	5.12	6.55	2.52	2.35	2.25	2.15	1.01	0.94	2.75	3.32	0.59	0.69
1988	154.34	203.79	55.04	75.57	8.18	17.46	4.02	6.59	1.47	2.35	1.66	2.12	0.75	0.94	2.87	3.19	0.31	0.69
1989	187.56	201.66	60.42	75.14	17.64	16.98	5.19	6.47	2.31	2.33	2.09	2.09	0.91	0.93	3.39	3.02	0.93	0.68
1990	186.31	197.76	47.36	74.29	18.64	16.60	5.15	6.34	2.11	2.30	2.41	2.05	1.13	0.92	2.75	2.84	0.67	0.68
1991	216.27	193.80	73.42	73.32	18.78	16.27	5.38	6.19	1.37	2.26	1.26	2.02	0.46	0.90	2.31	2.67	0.45	0.67
1992	137.13	189.70	39.85	72.28	12.70	15.98	3.77	6.05	1.07	2.22	0.89	1.99	0.44	0.88	1.74	2.51	0.35	0.66
1993	248.73	186.64	102.62	71.41	20.65	15.69	9.77	5.89	2.83	2.19	2.67	1.99	1.13	0.87	2.69	2.38	0.61	0.65
1994	251.60	182.64	111.48	70.81	20.06	15.56	8.71	5.84	2.93	2.17	3.39	1.98	1.27	0.85	2.77	2.27	0.85	0.65
1995	128.79	179.38	42.98	70.48	13.67	15.50	5.35	5.84	1.48	2.16	1.39	1.99	0.49	0.85	1.47	2.18	0.39	0.65
1996	238.82	176.82	153.66	70.41	25.26	15.53	11.54	5.89	2.55	2.16	2.94	2.01	1.14	0.84	2.99	2.10	1.00	0.65
1997	196.60	175.97	87.41	70.58	16.29	15.53	6.30	5.90	2.45	2.16	2.76	2.04	1.14	0.85	2.80	2.04	0.92	0.65
1998	207.46	174.71	125.21	70.58	26.86	15.61	10.39	5.98	3.23	2.17	3.58	2.06	1.35	0.85	2.35	1.99	0.78	0.65
1999	99.87	174.26	24.01	70.41	5.68	15.65	1.61	6.03	1.10	2.17	0.76	2.08	0.32	0.86	1.29	1.95	0.44	0.65
2000	149.39	173.78	41.51	69.96	9.24	15.74	3.14	6.12	1.49	2.17	1.33	2.09	0.57	0.86	1.75	1.92	0.76	0.66
2001	101.94	174.05	37.14	69.56	5.89	15.77	3.20	6.17	1.16	2.18	1.03	2.11	0.44	0.87	1.64	1.90	0.80	0.67
2002	114.10	173.61	16.76	69.17	2.58	15.96	0.76	6.29	0.37	2.19	0.27	2.12	0.11	0.88	0.92	1.88	0.20	0.67
2003	214.19	173.68	138.63	68.92	32.72	16.14	13.58	6.38	5.00	2.21	3.72	2.13	1.35	0.89	2.95	1.87	1.41	0.68
2004	382.52	173.70	104.48	68.79	19.87	16.35	9.58	6.47	3.29	2.23	3.11	2.15	1.41	0.90	2.80	1.85	0.88	0.69
2005	235.74	174.23	67.18	68.99	13.82	16.43	6.60	6.47	2.05	2.25	2.03	2.16	0.95	0.92	1.93	1.84	0.66	0.69
2006	211.69	173.45	43.92	69.23	10.25	16.52	4.69	6.42	1.06	2.25	1.29	2.16	0.73	0.93	1.86	1.82	0.63	0.70
2007	174.47	172.74	56.36	68.98	16.33	16.47	5.13	6.27	2.65	2.25	2.31	2.16	1.15	0.93	1.68	1.79	0.75	0.71
2008	175.29	170.65	52.91	68.07	5.32	16.41	3.63	6.11	1.10	2.24	0.94	2.15	0.60	0.92	1.47	1.76	0.45	0.71
2009	126.04	168.80	42.60	67.27	7.36	16.43	3.17	5.96	1.23	2.24	1.03	2.16	0.55	0.92	1.45	1.73	0.64	0.71
2010	140.50	163.97	78.96	67.04	20.81	16.80	6.89	5.90	2.91	2.25	2.44	2.21	1.27	0.91	2.20	1.70	0.34	0.62

Table 4–2.   Nitrate annual flux and flow-normalized annual flux, in tons per day, for the nine Chesapeake Bay River Input Monitoring Stations obtained using WRTDS.

[SUS, Susquehanna River at Conowingo, Md.; POT, Potomac River at Chain Bridge at Washington, D.C.; JAM, James River at Cartersville, Va.; RAP, Rappahannock River near Fredericksburg, Va.; APP, Appomattox River at Matoaca, Va.; PAM, Pamunkey River near Hanover, Va.; MAT, Mattaponi River near Beulahville, Va.; PAT, Patuxent River near Bowie, Md.; CHO, Choptank River near Greensboro, Md.; FN flux, flow-normalized flux]

Year	SUS Flux	SUS FN flux	POT Flux	POT FN flux	JAM Flux	JAM FN flux	RAP Flux	RAP FN flux	APP Flux	APP FN flux	PAM Flux	PAM FN flux	MAT Flux	MAT FN flux	PAT Flux	PAT FN flux	CHO Flux	CHO FN flux
1981	81.60	124.29	19.95	43.01	2.22	5.93	0.79	2.84	0.22	0.54	0.18	0.75	0.08	0.19	1.22	1.85	0.20	0.33
1982	118.51	126.10	37.81	43.37	5.52	5.85	2.26	2.83	0.50	0.56	0.56	0.71	0.15	0.19	1.46	1.91	0.30	0.33
1983	110.35	128.05	42.56	43.85	6.04	5.80	2.89	2.80	0.75	0.57	0.69	0.67	0.24	0.20	1.99	1.96	0.40	0.33
1984	173.42	129.88	67.96	44.29	8.47	5.73	4.82	2.77	0.94	0.58	1.09	0.64	0.33	0.21	2.47	2.00	0.48	0.34
1985	91.78	132.56	30.19	44.85	4.07	5.67	1.55	2.75	0.39	0.59	0.46	0.62	0.16	0.21	1.54	2.03	0.15	0.35
1986	138.47	134.62	43.28	45.31	4.71	5.60	2.36	2.73	0.53	0.60	0.55	0.62	0.19	0.21	1.49	2.04	0.27	0.36
1987	127.18	136.69	42.26	45.82	6.77	5.52	2.16	2.72	0.73	0.60	0.66	0.61	0.24	0.21	1.76	2.01	0.34	0.37
1988	106.68	137.97	35.09	46.46	2.95	5.44	1.93	2.71	0.40	0.59	0.50	0.61	0.19	0.21	1.82	1.94	0.20	0.38
1989	127.99	139.27	40.16	47.18	5.72	5.37	2.09	2.68	0.58	0.58	0.61	0.61	0.20	0.21	2.07	1.85	0.49	0.39
1990	133.44	139.31	34.82	47.66	6.64	5.31	2.53	2.64	0.57	0.57	0.74	0.61	0.26	0.21	1.76	1.75	0.41	0.39
1991	157.59	139.24	49.84	48.08	6.36	5.24	2.58	2.59	0.40	0.57	0.42	0.62	0.12	0.20	1.50	1.66	0.30	0.39
1992	101.73	138.48	29.18	48.40	4.21	5.16	1.80	2.53	0.31	0.56	0.33	0.63	0.11	0.20	1.19	1.58	0.24	0.40
1993	179.03	137.73	66.76	48.82	6.86	5.05	4.22	2.49	0.79	0.55	0.87	0.64	0.27	0.20	1.69	1.50	0.39	0.40
1994	183.11	135.54	74.33	49.03	6.50	4.93	3.60	2.46	0.81	0.54	1.05	0.65	0.29	0.19	1.72	1.43	0.49	0.40
1995	98.43	133.65	31.79	48.89	3.88	4.84	1.94	2.45	0.38	0.54	0.49	0.66	0.13	0.19	1.00	1.37	0.28	0.40
1996	175.09	132.11	93.85	48.26	6.87	4.78	4.54	2.46	0.65	0.54	0.92	0.66	0.25	0.19	1.82	1.31	0.60	0.41
1997	147.62	131.66	60.89	47.52	5.52	4.76	3.17	2.49	0.66	0.54	0.88	0.68	0.26	0.19	1.72	1.28	0.57	0.41
1998	157.37	130.48	78.55	46.59	7.96	4.76	4.17	2.53	0.93	0.55	1.14	0.68	0.32	0.19	1.48	1.25	0.48	0.41
1999	77.33	129.53	17.52	45.63	1.79	4.78	0.86	2.56	0.29	0.56	0.28	0.69	0.08	0.20	0.86	1.23	0.28	0.42
2000	113.36	128.29	29.73	44.63	3.18	4.81	1.67	2.59	0.41	0.56	0.52	0.70	0.15	0.20	1.16	1.21	0.49	0.42
2001	78.05	127.42	26.16	43.76	1.91	4.85	1.58	2.63	0.32	0.56	0.39	0.70	0.12	0.21	1.09	1.19	0.50	0.42
2002	84.92	125.75	11.24	42.85	0.76	4.90	0.41	2.66	0.11	0.56	0.11	0.69	0.03	0.21	0.65	1.17	0.15	0.43
2003	155.79	124.40	77.43	42.04	9.13	4.95	5.32	2.69	1.01	0.57	1.06	0.69	0.29	0.21	1.67	1.15	0.80	0.43
2004	202.50	123.17	63.27	41.29	6.77	5.00	4.31	2.72	0.77	0.57	0.94	0.68	0.32	0.22	1.66	1.14	0.56	0.44
2005	166.95	122.93	42.07	40.77	5.09	5.07	3.21	2.74	0.59	0.57	0.68	0.68	0.26	0.23	1.21	1.13	0.44	0.44
2006	146.10	122.20	29.26	40.30	3.42	5.10	2.43	2.71	0.32	0.58	0.47	0.67	0.20	0.23	1.13	1.11	0.43	0.45
2007	127.92	121.52	35.07	39.74	5.31	5.07	2.52	2.63	0.68	0.57	0.68	0.66	0.29	0.23	1.07	1.10	0.49	0.46
2008	128.97	119.90	31.80	38.72	1.92	4.93	1.37	2.51	0.26	0.55	0.29	0.64	0.14	0.23	0.92	1.08	0.31	0.46
2009	94.86	117.60	25.89	37.26	2.50	4.72	1.46	2.38	0.35	0.53	0.33	0.61	0.14	0.21	0.89	1.06	0.41	0.46
2010	102.52	112.77	39.84	35.48	5.54	4.50	2.72	2.24	0.74	0.52	0.66	0.58	0.28	0.20	1.35	1.03	0.70	0.47

Table 4–3.   Total phosphorus annual flux and flow-normalized annual flux, in tons per day, for the nine Chesapeake Bay River Input Monitoring Stations obtained using WRTDS.

[SUS, Susquehanna River at Conowingo, Md.; POT, Potomac River at Chain Bridge at Washington, D.C.; JAM, James River at Cartersville, Va.; RAP, Rappahannock River near Fredericksburg, Va.; APP, Appomattox River at Matoaca, Va.; PAM, Pamunkey River near Hanover, Va.; MAT, Mattaponi River near Beulahville, Va.; PAT, Patuxent River near Bowie, Md.; CHO, Choptank River near Greensboro, Md.; FN Flux, flow-normalized flux]

Year	SUS Flux	SUS FN flux	POT Flux	POT FN flux	JAM Flux	JAM FN flux	RAP Flux	RAP FN flux	APP Flux	APP FN flux	PAM Flux	PAM FN flux	MAT Flux	MAT FN flux	PAT Flux	PAT FN flux	CHO Flux	CHO FN flux
1981	6.217	9.534	2.030	7.207	1.087	3.649	0.087	0.673	0.041	0.210	0.046	0.204	0.026	0.076	0.322	0.507	0.016	0.044
1982	8.093	9.028	4.877	7.370	3.256	3.764	0.514	0.697	0.150	0.205	0.145	0.205	0.051	0.079	0.319	0.471	0.022	0.043
1983	7.613	8.673	6.154	7.407	3.757	3.870	1.000	0.708	0.249	0.205	0.200	0.204	0.099	0.080	0.401	0.436	0.052	0.042
1984	13.398	8.361	12.344	7.321	5.763	3.990	1.540	0.732	0.346	0.203	0.366	0.203	0.146	0.080	0.525	0.401	0.057	0.041
1985	4.378	8.135	2.674	6.984	2.426	4.097	0.256	0.759	0.098	0.201	0.147	0.203	0.056	0.081	0.218	0.367	0.010	0.041
1986	8.104	7.894	11.685	6.566	6.891	4.212	0.542	0.790	0.187	0.198	0.191	0.204	0.075	0.082	0.165	0.335	0.018	0.040
1987	6.250	7.631	4.337	6.204	6.485	4.279	0.709	0.841	0.220	0.198	0.212	0.206	0.089	0.084	0.222	0.305	0.029	0.040
1988	4.500	7.340	3.942	6.031	1.888	4.295	0.453	0.903	0.104	0.204	0.157	0.209	0.066	0.086	0.214	0.276	0.011	0.040
1989	6.730	7.067	4.016	5.854	4.267	4.244	0.882	0.933	0.196	0.214	0.207	0.215	0.085	0.090	0.280	0.247	0.055	0.039
1990	5.390	6.783	2.231	5.659	4.260	4.142	0.605	0.945	0.182	0.226	0.253	0.219	0.111	0.091	0.199	0.222	0.032	0.038
1991	7.008	6.537	4.715	5.477	4.422	3.970	0.695	0.933	0.124	0.234	0.133	0.221	0.042	0.090	0.151	0.199	0.019	0.038
1992	3.516	6.339	1.882	5.308	2.960	3.771	0.478	0.916	0.093	0.238	0.086	0.221	0.040	0.089	0.097	0.181	0.012	0.037
1993	11.003	6.254	7.707	5.092	4.637	3.567	1.460	0.879	0.318	0.239	0.297	0.221	0.110	0.087	0.183	0.167	0.027	0.037
1994	10.307	6.244	8.281	4.952	4.328	3.431	1.385	0.868	0.346	0.237	0.385	0.221	0.127	0.086	0.200	0.157	0.054	0.037
1995	3.454	6.292	2.233	4.931	3.066	3.328	1.147	0.860	0.129	0.234	0.145	0.223	0.046	0.085	0.077	0.152	0.015	0.038
1996	10.651	6.354	18.224	5.147	5.354	3.256	1.948	0.866	0.261	0.230	0.330	0.227	0.113	0.084	0.247	0.149	0.064	0.038
1997	7.564	6.439	6.133	5.488	3.215	3.182	0.736	0.862	0.253	0.225	0.328	0.232	0.115	0.083	0.221	0.147	0.059	0.039
1998	7.652	6.535	11.438	5.875	5.599	3.142	1.735	0.879	0.357	0.219	0.399	0.236	0.126	0.082	0.167	0.146	0.045	0.039
1999	2.817	6.690	1.157	6.208	1.060	3.104	0.161	0.893	0.100	0.215	0.091	0.239	0.029	0.081	0.088	0.146	0.032	0.040
2000	4.724	6.836	2.339	6.393	1.631	3.086	0.313	0.919	0.125	0.211	0.153	0.243	0.052	0.079	0.120	0.146	0.045	0.041
2001	2.865	6.997	2.097	6.441	1.060	3.070	0.364	0.936	0.102	0.209	0.122	0.249	0.039	0.079	0.111	0.148	0.053	0.042
2002	3.846	7.158	0.948	6.436	0.489	3.120	0.050	0.982	0.028	0.209	0.041	0.255	0.009	0.079	0.047	0.150	0.008	0.043
2003	8.937	7.353	15.831	6.382	7.001	3.198	2.317	1.030	0.499	0.210	0.472	0.262	0.124	0.079	0.308	0.153	0.114	0.043
2004	21.724	7.519	9.735	6.310	3.767	3.316	1.497	1.090	0.304	0.212	0.407	0.269	0.129	0.079	0.274	0.156	0.058	0.044
2005	11.863	7.652	5.091	6.205	2.365	3.418	1.021	1.147	0.185	0.214	0.251	0.276	0.081	0.079	0.159	0.158	0.037	0.045
2006	9.920	7.709	2.675	6.100	2.026	3.550	0.619	1.204	0.088	0.216	0.164	0.279	0.063	0.079	0.177	0.156	0.039	0.045
2007	6.952	7.771	3.739	5.920	3.361	3.668	0.784	1.271	0.269	0.219	0.304	0.283	0.098	0.079	0.134	0.153	0.045	0.045
2008	7.149	7.854	3.722	5.774	0.778	3.827	0.864	1.334	0.099	0.222	0.123	0.290	0.053	0.079	0.122	0.149	0.024	0.045
2009	3.778	8.126	2.790	5.803	1.101	4.100	0.483	1.399	0.114	0.227	0.134	0.305	0.048	0.080	0.123	0.147	0.046	0.046
2010	5.549	8.287	7.078	6.117	5.542	4.483	1.970	1.516	0.331	0.233	0.346	0.326	0.111	0.082	0.194	0.148	0.098	0.047

Table 4–4. Orthophosphorus annual flux and flow-normalized annual flux, in tons per day, for the nine Chesapeake Bay River Input Monitoring Stations obtained using WRTDS.

[SUS, Susquehanna River at Conowingo, Md.; POT, Potomac River at Chain Bridge at Washington, D.C.; JAM, James River at Cartersville, Va.; RAP, Rappahannock River near Fredericksburg, Va.; APP, Appomattox River at Matoaca, Va.; PAM, Pamunkey River near Hanover, Va.; MAT, Mattaponi River near Beulahville, Va.; PAT, Patuxent River near Bowie, Md.; CHO, Choptank River near Greensboro, Md.; FN flux, flow-normalized flux]

Year	SUS Flux	SUS FN flux	POT Flux	POT FN flux	JAM Flux	JAM FN flux	RAP Flux	RAP FN flux	APP Flux	APP FN flux	PAM Flux	PAM FN flux	MAT Flux	MAT FN flux	PAT Flux	PAT FN flux	CHO Flux	CHO FN flux
1982	1.627	1.803	1.003	1.281	0.916	0.946	0.063	0.097	0.032	0.039	0.035	0.046	0.017	0.022	0.299	0.340	0.008	0.013
1983	1.224	1.639	1.215	1.312	1.087	1.057	0.083	0.100	0.054	0.039	0.047	0.049	0.029	0.022	0.215	0.228	0.015	0.011
1984	1.834	1.468	2.229	1.335	1.464	1.174	0.171	0.103	0.066	0.038	0.087	0.050	0.037	0.022	0.180	0.158	0.015	0.010
1985	0.821	1.314	0.751	1.343	1.149	1.260	0.051	0.103	0.024	0.038	0.038	0.051	0.016	0.021	0.095	0.113	0.002	0.010
1986	1.189	1.187	1.275	1.332	1.171	1.306	0.112	0.101	0.037	0.037	0.048	0.050	0.019	0.020	0.067	0.085	0.005	0.009
1987	1.046	1.098	1.116	1.316	1.435	1.327	0.075	0.096	0.041	0.037	0.048	0.048	0.022	0.019	0.062	0.069	0.006	0.009
1988	0.754	1.044	0.936	1.297	1.069	1.327	0.059	0.091	0.027	0.037	0.038	0.046	0.016	0.019	0.055	0.060	0.003	0.010
1989	0.907	1.019	1.173	1.262	1.347	1.295	0.068	0.087	0.039	0.038	0.045	0.045	0.019	0.018	0.058	0.054	0.016	0.010
1990	0.949	1.012	0.797	1.216	1.411	1.228	0.076	0.085	0.039	0.038	0.052	0.045	0.023	0.018	0.049	0.049	0.010	0.010
1991	1.286	1.024	1.140	1.176	1.265	1.140	0.082	0.084	0.026	0.040	0.032	0.045	0.010	0.018	0.040	0.045	0.006	0.011
1992	0.690	1.040	0.583	1.148	0.951	1.041	0.054	0.083	0.022	0.041	0.026	0.046	0.011	0.018	0.031	0.042	0.005	0.011
1993	1.405	1.062	1.516	1.124	1.088	0.946	0.138	0.082	0.054	0.041	0.057	0.047	0.023	0.019	0.041	0.039	0.009	0.011
1994	1.511	1.077	1.725	1.109	0.971	0.867	0.125	0.082	0.058	0.041	0.074	0.049	0.026	0.019	0.043	0.037	0.016	0.011
1995	0.766	1.098	0.678	1.104	0.743	0.805	0.072	0.083	0.029	0.041	0.041	0.051	0.013	0.019	0.024	0.035	0.005	0.011
1996	1.470	1.120	2.921	1.106	0.944	0.756	0.171	0.084	0.051	0.040	0.073	0.053	0.027	0.019	0.052	0.034	0.019	0.011
1997	1.364	1.144	1.493	1.115	0.784	0.718	0.096	0.084	0.049	0.040	0.071	0.056	0.026	0.019	0.046	0.032	0.016	0.011
1998	1.384	1.151	1.915	1.125	0.880	0.684	0.150	0.086	0.062	0.039	0.083	0.058	0.027	0.019	0.033	0.030	0.012	0.011
1999	0.621	1.144	0.348	1.123	0.417	0.649	0.023	0.087	0.020	0.038	0.033	0.059	0.007	0.018	0.021	0.028	0.012	0.011
2000	0.924	1.129	0.673	1.104	0.522	0.603	0.049	0.088	0.027	0.037	0.050	0.060	0.013	0.017	0.024	0.026	0.012	0.011
2001	0.592	1.114	0.530	1.072	0.366	0.552	0.047	0.088	0.020	0.036	0.041	0.059	0.009	0.016	0.021	0.024	0.013	0.011
2002	0.639	1.089	0.227	1.033	0.243	0.500	0.011	0.090	0.006	0.035	0.024	0.057	0.003	0.016	0.013	0.023	0.002	0.010
2003	1.436	1.073	2.234	0.986	0.750	0.452	0.209	0.091	0.080	0.035	0.079	0.055	0.023	0.015	0.036	0.021	0.028	0.010
2004	2.067	1.073	1.637	0.932	0.508	0.409	0.152	0.092	0.052	0.034	0.071	0.052	0.022	0.014	0.031	0.020	0.013	0.010
2005	1.603	1.087	0.864	0.870	0.362	0.368	0.091	0.093	0.032	0.033	0.049	0.049	0.014	0.013	0.020	0.019	0.009	0.011
2006	1.423	1.088	0.542	0.804	0.248	0.325	0.074	0.093	0.018	0.033	0.036	0.045	0.010	0.012	0.022	0.018	0.011	0.011
2007	1.179	1.084	0.586	0.744	0.275	0.278	0.072	0.092	0.041	0.033	0.044	0.042	0.014	0.012	0.017	0.018	0.011	0.012
2008	1.105	1.080	0.506	0.701	0.124	0.233	0.054	0.090	0.017	0.032	0.022	0.038	0.008	0.011	0.015	0.017	0.007	0.012
2009	0.784	1.084	0.438	0.690	0.135	0.197	0.048	0.088	0.019	0.032	0.021	0.034	0.007	0.011	0.016	0.017	0.017	0.013
2010	0.931	1.089	0.613	0.699	0.183	0.174	0.063	0.097	0.041	0.031	0.033	0.032	0.013	0.010	0.021	0.017	0.024	0.014

Table 4–5.   Suspended sediment annual flux and flow-normalized annual flux, in tons per day, for the nine Chesapeake Bay River Input Monitoring Stations obtained using WRTDS.

[SUS, Susquehanna River at Conowingo, Md.; POT, Potomac River at Chain Bridge at Washington, D.C.; JAM, James River at Cartersville, Va.; RAP, Rappahannock River near Fredericksburg, Va.; APP, Appomattox River at Matoaca, Va.; PAM, Pamunkey River near Hanover, Va.; MAT, Mattaponi River near Beulahville, Va.; PAT, Patuxent River near Bowie, Md.; CHO, Choptank River near Greensboro, Md.; FN flux, flow-normalized flux; NA, not available]

Year	SUS Flux	SUS FN flux	POT Flux	POT FN flux	JAM Flux	JAM FN flux	RAP Flux	RAP FN flux	APP Flux	APP FN flux	PAM Flux	PAM FN flux	MAT Flux	MAT FN flux	PAT Flux	PAT FN flux	CHO Flux	CHO FN flux
1981	3,621.21	5,525.57	1,081.69	6,423.90	NA	NA	NA	NA	NA	NA	NA	NA	NA	NA	19.24	76.94	2.72	12.50
1982	4,106.13	5,142.80	4,206.75	6,838.32	NA	NA	NA	NA	NA	NA	NA	NA	NA	NA	34.61	82.01	3.85	11.40
1983	4,093.29	4,773.36	6,678.15	7,194.09	NA	NA	NA	NA	NA	NA	NA	NA	NA	NA	92.91	87.73	14.87	10.69
1984	9,891.60	4,418.14	13,712.10	7,296.24	NA	NA	NA	NA	NA	NA	NA	NA	NA	NA	139.12	93.44	14.85	10.25
1985	1,377.63	4,137.09	1,714.13	6,813.71	NA	NA	NA	NA	NA	NA	NA	NA	NA	NA	34.67	98.16	1.71	10.05
1986	3,979.10	3,903.02	11,264.56	6,142.33	NA	NA	NA	NA	NA	NA	NA	NA	NA	NA	23.31	101.13	3.53	9.96
1987	2,315.77	3,738.22	3,942.48	5,596.20	NA	NA	NA	NA	NA	NA	NA	NA	NA	NA	58.83	101.18	6.82	9.93
1988	1,481.33	3,647.11	3,524.21	5,306.95	NA	NA	NA	NA	NA	NA	NA	NA	NA	NA	65.24	98.05	2.00	9.83
1989	2,956.75	3,620.06	3,584.16	5,142.96	NA	NA	NA	NA	NA	NA	NA	NA	NA	NA	129.07	92.60	12.46	9.66
1990	1,860.09	3,608.91	1,297.55	4,710.71	NA	NA	NA	NA	NA	NA	NA	NA	NA	NA	71.78	86.42	6.50	9.21
1991	3,181.82	3,616.11	3,008.85	4,271.05	NA	NA	NA	NA	NA	NA	NA	NA	NA	NA	47.41	80.39	3.52	8.74
1992	1,224.68	3,617.88	1,107.94	3,894.95	NA	NA	NA	NA	NA	NA	NA	NA	NA	NA	28.18	75.13	1.86	8.27
1993	9,014.61	3,626.38	6,204.80	3,594.03	NA	NA	NA	NA	NA	NA	NA	NA	NA	NA	81.92	71.08	6.08	7.88
1994	6,913.26	3,617.25	5,641.09	3,357.52	NA	NA	NA	NA	NA	NA	NA	NA	NA	NA	96.16	68.62	14.08	7.55
1995	1,276.31	3,609.81	1,087.07	3,243.40	NA	NA	NA	NA	NA	NA	NA	NA	NA	NA	22.38	67.23	2.35	7.26
1996	10,244.86	3,631.89	15,535.52	3,215.35	NA	NA	NA	NA	NA	NA	NA	NA	NA	NA	125.46	66.66	10.69	7.02
1997	4,305.70	3,715.54	2,691.54	3,224.47	NA	NA	NA	NA	NA	NA	NA	NA	NA	NA	103.29	66.35	11.69	6.77
1998	4,206.36	3,832.61	6,737.45	3,350.07	NA	NA	NA	NA	NA	NA	NA	NA	NA	NA	82.53	66.36	9.72	6.49
1999	1,099.71	3,992.14	313.81	3,545.30	NA	NA	NA	NA	NA	NA	NA	NA	NA	NA	34.82	66.29	3.86	6.30
2000	1,785.41	4,205.54	665.94	3,792.24	978.05	2,261.91	NA	NA	31.00	51.34	79.23	164.15	13.31	30.65	46.44	66.47	6.22	6.18
2001	909.92	4,509.06	685.86	4,041.54	696.45	2,042.28	379.04	936.95	28.58	52.72	66.83	143.54	9.83	25.40	43.07	66.99	7.51	6.18
2002	1,451.45	4,866.30	304.76	4,278.02	157.93	2,039.98	27.03	905.06	5.03	53.67	5.49	131.63	1.33	22.42	10.25	68.06	0.67	6.23
2003	4,069.15	5,202.07	12,073.42	4,503.57	7,561.96	2,025.57	2,630.80	892.40	175.58	54.52	347.24	126.09	35.46	20.78	168.16	69.35	16.16	6.28
2004	34,721.41	5,412.79	6,291.05	4,745.19	2,752.18	2,078.46	1,389.33	889.37	89.96	55.19	237.88	124.46	35.49	19.68	129.76	70.75	8.73	6.37
2005	8,339.01	5,598.06	3,207.44	5,014.75	1,450.07	1,972.78	892.37	883.43	47.05	55.92	129.19	124.91	21.30	18.93	71.55	71.71	5.19	6.48
2006	7,567.06	5,670.85	1,394.72	5,374.35	1,702.64	2,041.59	448.00	888.72	18.81	56.32	65.00	127.14	14.25	18.22	81.97	71.83	4.75	6.63
2007	3,139.77	6,082.02	2,652.55	5,783.14	2,548.40	2,064.14	554.62	894.00	79.53	56.66	167.79	131.36	24.88	17.70	59.36	71.51	7.41	6.79
2008	4,017.29	6,396.86	3,562.38	6,337.82	398.59	2,172.14	799.26	906.69	22.35	56.85	69.88	139.16	12.29	17.36	60.48	71.81	3.49	7.05
2009	1,198.85	7,349.21	2,589.70	6,929.49	621.42	2,134.88	307.63	922.76	25.33	57.34	73.44	153.08	10.84	17.48	64.05	73.80	5.61	7.42
2010	3,125.31	7,714.94	9,802.96	7,642.59	4,359.82	2,333.74	1,737.02	959.29	93.40	58.05	232.88	176.64	27.41	18.10	100.54	77.77	19.58	7.79